Selected Titles in This Series

Volume

11 **Spencer J. Bloch**
Higher regulators, algebraic K-theory, and zeta functions of elliptic curves
2000

10 **James D. Lewis**
A survey of the Hodge conjecture, Second Edition
1999

9 **Yves Meyer**
Wavelets, vibrations and scaling
1998

8 **Ioannis Karatzas**
Lectures on the mathematics of finance
1996

7 **John Milton**
Dynamics of small neural populations
1996

6 **Eugene B. Dynkin**
An introduction to branching measure-valued processes
1994

5 **Andrew Bruckner**
Differentiation of real functions
1994

4 **David Ruelle**
Dynamical zeta functions for piecewise monotone maps of the interval
1994

3 **V. Kumar Murty**
Introduction to Abelian varieties
1993

2 **M. Ya. Antimirov, A. A. Kolyshkin, and Rémi Vaillancourt**
Applied integral transforms
1993

1 **D. V. Voiculescu, K. J. Dykema, and A. Nica**
Free random variables
1992

Volume 11

CRM MONOGRAPH SERIES

Centre de Recherches Mathématiques
Université de Montréal

Higher Regulators, Algebraic *K*-Theory, and Zeta Functions of Elliptic Curves

Spencer J. Bloch

The Centre de Recherches Mathématiques (CRM) of the Université de Montréal was created in 1968 to promote research in pure and applied mathematics and related disciplines. Among its activities are special theme years, summer schools, workshops, postdoctoral programs, and publishing. The CRM is supported by the Université de Montréal, the Province of Québec (FCAR), and the Natural Sciences and Engineering Research Council of Canada. It is affiliated with the Institut des Sciences Mathématiques (ISM) of Montréal, whose constituent members are Concordia University, McGill University, the Université de Montréal, the Université du Québec à Montréal, and the Ecole Polytechnique.

American Mathematical Society
Providence, Rhode Island USA

The production of this volume was supported in part by the Fonds pour la Formation de Chercheurs et l'Aide à la Recherche (Fonds FCAR) and the Natural Sciences and Engineering Research Council of Canada (NSERC).

2000 *Mathematics Subject Classification.* Primary 19F27; Secondary 14G10, 19D50.

ABSTRACT. Work of Borel on higher regulators for number fields is discussed. The Borel regulator for K_3 of a number field is described explicitly in terms of the dilogarithm function.

A generalization, based on functions related to the dilogarithm and to the Dedekind η-function, leads to a regulator for K_2 of an elliptic curve E over a number field. Elements in $K_2(E)$ analogous to cyclotomic units are described. The regulator is evaluated on these elements and the resulting values related to the value of the Hasse-Weil zeta function of E at $s = 2$ when E has complex multiplication. This regulator formula is worked out in detail for the case of E defined over $\mathbb{Q}$ with complex multiplication by the ring of integers in an imaginary quadratic field, when it takes a particularly simple form.

Library of Congress Cataloging-in-Publication Data

Bloch, Spencer.

Higher regulators, algebraic K-theory, and zeta functions for elliptic curves / Spencer J. Bloch.

p. cm. — (CRM monograph series, ISSN 1065-8599 ; v. 11)

Includes bibliographical references and index.

ISBN 0-8218-2114-8 (alk. paper)

1. Curves, Elliptic. 2. K-theory. 3. Functions, Zeta. I. Title. II. Series.

QA567.2.E44.B56 2000

516.3′52—dc21 00-029969

∞ The paper used in this book is acid-free and falls within the guidelines
established to ensure permanence and durability.
This volume was typeset using AMS-TeX,
the American Mathematical Society's TeX macro system,
and submitted to the American Mathematical Society in camera ready
form by the Centre de Recherches Mathématiques.
Visit the AMS home page at URL: http://www.ams.org/

10 9 8 7 6 5 4 3 2 1 05 04 03 02 01 00

Author's Apology

"The author apologizes for the long delay in publishing this monograph. The reader should understand that since this work was done, fundamental ideas of A. Beilinson, A. Suslin, V. Voevodsky, and others have totally transformed the landscape. Sometimes it is fun to drive around in a Model T Ford but one should be aware there are much faster cars on the road."

Contents

Acknowledgments

I should like to express my gratitude to Bill Messing and the Department of Mathematics at Irvine for inviting me to give these lectures. At the time of the lectures the whole relationship between the regulator and the zeta function of an elliptic curve was purely conjectural. Messing's interest gave me the fortitude to push through the complicated business. I also want to acknowledge many fruitful conversations with D. Wigner about the dilogarithm function. It was he who first wrote down the function $D(x)$ (0.3.4) and worked out a number of its properties.

Spencer J. Bloch
Department of Mathematics
The University of Chicago
Chicago, IL 60637
USA

LECTURE 0

Introduction

0.1. Let K be a number field with ring of integers $\mathcal{O}_K$. Let

$$\mathcal{O}_K \otimes_{\mathbb{Z}} \mathbb{R} = \mathbb{R}^{r_1} \oplus \mathbb{C}^{r_2},$$

so r_1 and r_2 are the numbers of real and complex places of K, and define

$$\ell = (\ell_1, \ldots, \ell_{r_1+r_2})\colon \mathcal{O}_K^* \longrightarrow \mathbb{R}^{r_1+r_2} \tag{0.1.1}$$

to be the composite of the inclusion $\mathcal{O}_K^* \hookrightarrow (\mathcal{O}_K \otimes_{\mathbb{Z}} \mathbb{R})^*$ with the map $(\mathcal{O}_K \otimes \mathbb{R})^* \longrightarrow \mathbb{R}^{r_1+r_2}$ given by $\log|\ |$ (resp. $\log|\ |^2$) on the real (resp. complex) factors. The image $\ell(\mathcal{O}_K^*) = L$ is known to be a lattice of rank $r_1 + r_2 - 1$, and the *regulator* R_K is defined by

$$R_K = \frac{1}{\sqrt{r_1 + r_2}} \operatorname{Volume}(L). \tag{0.1.2}$$

If $\epsilon_1, \ldots, \epsilon_{r_1+r_2-1}$ is a basis of $\mathcal{O}_K^*/\text{torsion}$, it is easy to show R_K is the absolute value of the determinant of any $(r_1 + r_2 - 1) \times (r_1 + r_2 - 1)$ minor of the matrix $\big(\ell_i(\epsilon_j)\big)_{1 \le i,j \le r_1+r_2}$. The *class number formula*

$$\lim_{s\to 1}(s-1)\zeta_K(s) = \frac{2^{r_1+r_2}\pi^{r_2} R_K h}{\sqrt{|D|}w} \tag{0.1.3}$$

relates the residue at $s = 1$ of the zeta function $\zeta_K(s)$ to the regulator R_K, the class number h, the number w of roots of 1, and the discriminant D of K. If one takes into account the functional equation satisfied by $\zeta_K(s)$, the class number formula can be rewritten ([**Lic73**])

$$\lim_{s\to 0} \zeta_K(s) s^{-(r_1+r_2-1)} = \frac{-hR_K}{w}. \tag{0.1.4}$$

Various conjectural generalizations of the class number formula have been proposed. One such, due to Lichtenbaum, would have

$$\lim_{s\to -m} \zeta_K(s)(s+m)^{-d_m} = \pm \frac{\# K_{2m}(\mathcal{O}_K)}{\# K_{2m+1}(\mathcal{O}_K)_{\text{tor}}} R_m(K) \tag{0.1.5}$$

for any non-negative integer m, where $K_*(\mathcal{O}_K)$ denotes the K-theory of Bass, Milnor, and Quillen, among others, $R_m(K)$ is a suitable regulator, and

$$d_m = \begin{cases} r_2 & m = 2n+1 > 0 \\ r_1 + r_2 & m = 2n > 0 \\ r_1 + r_2 - 1 & m = 0. \end{cases} \tag{0.1.6}$$

0.2. Unfortunately, Lichtenbaum's conjecture does not give the right value, even for $m = 1$ and $K = \mathbb{Q}$. However, various results suggest some formula of this sort should hold. In particular Borel has computed the rank of the $K_*(\mathcal{O}_K)$, proving

$$\operatorname{rk} K_n(\mathcal{O}_K) = \begin{cases} 0 & n = 2m \\ d_m & n = 2m+1. \end{cases} \tag{0.2.1}$$

So $\operatorname{rk} K_{2m+1}(\mathcal{O}_K) =$ order of zero of $\zeta_K(s)$ at $s = -m$. The indecomposable space I^{2m+1} of the continuous cohomology

$$H^{2m+1}_{\mathrm{cont}}(\mathrm{SL}(\mathcal{O}_K \otimes_{\mathbb{Z}} \mathbb{R}), \mathbb{R})$$

has dimension d_m, and the Hurewitz map

$$K_*(\mathcal{O}_K) \to H_*\big(\mathrm{GL}(\mathcal{O}_K)\big)$$

gives a homomorphism

$$r\colon K_*(\mathcal{O}_K) \to \mathrm{Hom}(I^{2m+1}, \mathbb{R}). \tag{0.2.2}$$

Borel shows further that r embeds $K_*(\mathcal{O}_K)/\text{torsion}$ as a lattice of maximal rank, and that for a suitable choice of basis (obtained by relating $H^*_{\mathrm{cont}}(\mathrm{SL}(\mathcal{O}_K \otimes \mathbb{R}), \mathbb{R})$ to the cohomology of the compact dual symmetric space of $\mathrm{SL}(\mathcal{O}_K \otimes \mathbb{R})$), $r\big(K_{2m+1}(\mathcal{O}_K)\big)$ has volume some rational multiple of

$$\pi^{-d_m} \lim_{s \to -m} \zeta_K(s)(s+m)^{-d_m} = \pi^{-d(m+1)} |D|^{1/2} \zeta_K(m+1). \tag{0.2.3}$$

The first four lectures are devoted to Borel's work, although we completely neglect the more difficult aspect, the computation of $\operatorname{rk} K_*(\mathcal{O}_K)$, in order to focus on the adelic Tamagawa-Weil techniques for determining the volume of $r\big(K_*(\mathcal{O}_K)\big)$. Lectures 1 and 2 recall necessary ideas from Weil's Princeton notes [**Wei61**], while Lectures 3 and 4 focus on Borel's preprint [**Bor77**].

0.3. The remaining lectures are more tentative, and the results are much less sweeping. They represent the author's attempt to extend Borel type regulator results from number fields to elliptic curves over number fields. After all, Quillen has defined higher K-groups $K_*(X)$

for any scheme X. One might hope for a computation of the rank d of $K_*(X)$ together with a regulator map

$$r\colon K_*(X) \to \mathbb{R}^d$$

such that $r(K_*(X))$ was a lattice of maximal rank whose volume was related to the value of the Hasse-Weil zeta function of X. Unfortunately, such results do not come by magic, and it should be pointed out that appropriate generalizations of the two basic tools in Borel's arguments (viz., symmetric spaces and idèles) are not at this time available.

A conjectural generalization of the class number formula, due to Birch and Swinnerton-Dyer has impressive numerical evidence behind it. The idea is to describe the behavior of the Hasse-Weil zeta function of an elliptic curve E over a number field K near the point $s = 1$. The role of the regulator is played by the matrix associated to the height pairing on points of the curve defined over the given number field. In particular, one expects that the order of the zero of the Hasse-Weil zeta function at $s = 1$ should equal the rank of $E(K)$. Recently Coates-Wiles proved that when E has complex multiplication by the ring of integers in an imaginary quadratic number field k of class number 1, and when $K = \mathbb{Q}$ or k then the existence of points of infinite order in $E(K)$ implies the vanishing of the Hasse-Weil zeta function at $s = 1$. No one has been able to directly relate the height pairing matrix with the zeta function.

Oddly enough, if one looks at $s = 2$ rather than $s = 1$, things get easier. For one thing, instead of $K_0(E)$ or $E(K)$, the regulator will involve $K_2(E)$. It turns out to be possible to:

(a) write down explicitly a regulator map $K_2(E_{\mathbb{C}}) \to \mathbb{C}$ (not $\mathbb{R}$!);
(b) write down interesting elements in $K_2(E_K)$ analogous to cyclotomic units and related to points of finite order on E;
(c) evaluate the regulator map on the interesting elements and relate the resulting mess to $L(E, s)$, the Hasse-Weil zeta function, assuming E has complex multiplication.

The simplest case is when E is defined over $\mathbb{Q}$ and has complex multiplication by the ring of integers in an imaginary quadratic field (necessarily of class number 1 since $j(E) \in \mathbb{Q}$ generates the Hilbert class field). In the last lecture we write down an element $U \in K_2(E) \otimes_{\mathbb{Z}} \mathbb{Q}$, and show

$$\kappa R_q(U) = L(E, 2) \tag{0.3.1}$$

where R_q is the regulator (which depends "modularly" on the choice of $q = e^{2\pi i \tau}$) and κ is a certain constant involving a Gauss sum, the imaginary part of τ, and the conductor of E.

Some comments on (a). The construction is purely transcendental and we may think of E as defined over $\mathbb{C}$. We define a bilinear map

$$(0.3.2) \qquad \begin{aligned} \mathbb{C}(E)^* \otimes_{\mathbb{Z}} \mathbb{C}(E)^* &\to \textstyle\coprod_{p\in E} \mathbb{Z} \\ f \otimes g &\longmapsto (f)^- * (g) \end{aligned}$$

where (f) denotes the divisor of f, $\left(\sum n_i(a_i)\right)^- = \sum n_i(-a_i)$, and

$$\sum n_i(a_i) * \sum m_j(b_j) = \sum n_i m_j(a_i + b_j).$$

Thus any set-theoretic function $F\colon E_{\mathbb{C}} \to A$, A an Abelian group, induces a map $F^*\colon \mathbb{C}(E)^* \otimes \mathbb{C}(E)^* \to A$. It is natural to look for F such that F^* is a Steinberg symbol, i.e. $F^*\big(f \otimes (1-f)\big) = 0$ for any f. Such an F^* factors through $K_2\big(\mathbb{C}(E)\big)$ and we obtain by composition

$$(0.3.3) \qquad K_2(E) \to K_2\big(\mathbb{C}(E)\big) \xrightarrow{F^*} A.$$

Actually, it is necessary to back off a step and first consider the number field case. This means replacing $K_2(E_{\mathbb{C}})$ by $K_3(\mathbb{C})$ and writing down an explicit recipe for the Borel regulator, which is done in Lectures 5–7. We view $K_3(\mathbb{C})$ as a direct summand of the relative K-group $K_2(\mathbb{P}^1_{\mathbb{C}}, \{0,\infty\})$ which can be thought of either as K_2 of a degenerate elliptic curve or as K_2 with compact supports of $\mathbb{G}_m$. The analogue of $\mathbb{C}(E)$ becomes the semi-local ring R of functions on $\mathbb{P}^1_{\mathbb{C}}$ regular at 0 and ∞, and $K_2\big(\mathbb{C}(E)\big)$ is replaced by $K_2(R,I)$, $I =$ ideal of $\{0,\infty\}$. The group structure on $\mathbb{P}^1_{\mathbb{C}} \setminus \{0,\infty\}$ enables us to define

$$(f)^- * (g) \in \coprod_{p\in\mathbb{P}^1_{\mathbb{C}}\setminus\{0,\infty\}} \mathbb{Z}$$

for $f \in 1 + I$, $g \in \mathbb{C}(\mathbb{P}^1)^*$.

The function $D\colon \mathbb{P}^1_{\mathbb{C}} \to \mathbb{R}$,

$$(0.3.4) \qquad D(x) = \arg(1-x)\log|x| - \operatorname{Im}\int_0^x \log(1-t)\frac{dt}{t}$$

was first written down by D. Wigner, who showed it was continuous and single-valued, vanished at 0, 1, ∞, and represented an $\mathbb{R}$-valued continuous 3-cohomology class for the group $\mathrm{SL}_2(\mathbb{C})$. We show

$$(0.3.5) \qquad D\big((f)^- * (1-f)\big) = 0$$

for $f \in 1 + I$. Results of Keune on generators and relations for relative K_2 imply that D induces a map $K_2(R,I) \to \mathbb{R}$, and we may consider the composition (also denoted D)

$$K_3(\mathbb{C}) \to K_2(R,I) \to \mathbb{R}.$$

In Lecture 7 we do some number theoretic computations in the spirit of Kummer for cyclotomic fields, using D.

Returning to the elliptic case, we write $E = E_{\mathbb{C}} = \mathbb{C}^*/q^{\mathbb{Z}}$, $q = e^{2\pi i\tau}$, $\operatorname{Im}\tau > 0$. Using the relation $D(x^{-1}) = -D(x)$, we show that the infinite sum

$$\sum_{n\in\mathbb{Z}} D(xq^n) \overset{\text{dfn}}{=} D_q(x) \tag{0.3.6}$$

induces a continuous function $D_q\colon E \to \mathbb{R}$. In Lecture 9, we show

$$D_q\big((f)^- * (1-f)\big) = 0$$

for any elliptic function f, so D_q induces a map

$$\underbrace{K_2(E) \longrightarrow K_2\big(\mathbb{C}(E)\big) \longrightarrow}_{D_q} \mathbb{R}.$$

We may also consider the function $J\colon \mathbb{C} \to \mathbb{R}$,

$$J(x) = \log|x| \cdot \log|1-x|.$$

One checks $J\big((f)^- * (1-f)\big) = 0$ for $f \in 1 + I$, but the resulting map $J\colon K_2(R, I) \to \mathbb{R}$ appears less interesting than D because $K_3(\mathbb{C})$ maps to $\operatorname{Ker} J \subset K_2(R, I)$. Also the function

$$J_q = \sum_{n=0}^{\infty} J(xq^n) - \sum_{n=1}^{\infty} J(x^{-1}q^n) \tag{0.3.7}$$

is well-defined and continuous on $\mathbb{C}^*$, but is not invariant under $x \mapsto xq$. Given elliptic functions f, g, however, we may choose liftings $(\tilde{f})$, $(\tilde{g})$ to divisors on $\mathbb{C}^*$. It turns out that

$$J_q(f \otimes g) \overset{\text{dfn}}{=} J_q\big((\tilde{f})^- * (\tilde{g})\big)$$

is independent of liftings and $J_q\big(f \otimes (1-f)\big) = 0$. The regulator map is

$$R_q \overset{\text{dfn}}{=} J_q + \sqrt{-1}D_q. \tag{0.3.8}$$

Comments on (b); *interesting elements in* $K_2(E)$. When E is defined over K, there is an exact localization sequence

$$K_2(E) \to K_2\big(K(E)\big) \xrightarrow[\text{tame symbol}]{} \coprod_{x\in E} K(x)^*.$$

Suppose the points of order N, E_N, are defined over K, and let f and g be functions on E defined over K whose divisors are supported on E_N. Then one shows there is an expression

$$\{f, g\}^N \cdot \prod \{f_i, c_i\}, \quad f_i \in K(E)^*, c_i \in K^*$$

lying in the kernel of the tame symbol. The regulator map R_q is trivial on symbols with one entry constant, so it is possible to calculate its value on such an expression without specifying the f_i or the c_i. Let f be the function with poles of order 1 at each non-zero point of E_N and a zero of order $N^2 - 1$ at 0. Take $g = g_a$ to have a zero of order N at $a \in E_N$ and a pole of order N at 0. The resulting element in Image $\big(K_2(E) \to K_2\big(K(E)\big)\big)$ is denoted S_a. When K is a number field, the kernel of $K_2(E) \to K_2\big(K(E)\big)$ can be shown to be torsion, so $S_a \in K_2(E)/\text{torsion}$.

Comments on (c): *relation with* $L(E,2)$. The idea is to imitate the following classical computation of Kummer. Let

$$L(s,\chi) = \sum_{n=1}^{\infty} \frac{\chi(n)}{n^s}$$

be a Dirichlet L-series with conductor ℓ. Write

$$\widehat{\chi}(k) = \frac{1}{\ell} \sum_{r=0}^{\ell-1} \chi(r) e^{2\pi i r k/\ell}.$$

It is easy to see

$$L(s,\chi) = \sum_{k=0}^{\ell-1} \widehat{\chi}(k) \sum_{n=1}^{\infty} \frac{e^{-2\pi i n k/\ell}}{n^s}$$

so if χ is non-trivial we can write

$$L(1,\chi) = -\sum_{k=0}^{\ell-1} \widehat{\chi}(k) \log \left| \frac{1 - e^{-2\pi i k/\ell}}{1 - e^{2\pi i/\ell}} \right|$$

The point is that $(1 - e^{-2\pi i k/\ell})/(1 - e^{-2\pi i/\ell})$ is a unit in $\mathcal{O}_{\mathbb{Q}(e^{2\pi i/\ell})}$, and such units generate a subgroup of finite index in the full group of units. Since $\zeta_{\mathbb{Q}(e^{2\pi i/\ell})}(s)$ is a product of the $L(s,\chi)$, the above formula expresses the relation between the regulator and the residue of the zeta function at $s = 1$. One does not see the class number; it appears as the index of the above cyclotomic units in the full group of units.

The picture becomes even simpler if we consider a real quadratic extension K of $\mathbb{Q}$, and let χ be its character. Then $\lim_{s\to 1} \zeta_K(s) = L(1,\chi)$. The character χ is even and

$$\widehat{\chi}(k) = \chi(k)^{-1} \widehat{\chi}(1)$$

so we find

$$\lim_{s\to 1} \zeta_K(s)(s-1) = -\widehat{\chi}(1) \log |U| \tag{0.3.9}$$

where

$$U = \prod_{k \in (\mathbb{Z}/\ell\mathbb{Z})^*} \left(\frac{1 - e^{-2\pi i k/\ell}}{1 - e^{-2\pi i/\ell}} \right)^{\chi(k)^{-1}} \in \mathcal{O}_K^*$$

is the norm of a unit

$$\frac{1 - e^{-2\pi i k/\ell}}{1 - e^{-2\pi i/\ell}} \in \mathcal{O}^*_{\mathbb{Q}(e^{2\pi i/\ell})}.$$

This formula should be compared with the formula (0.3.1) above.

LECTURE 1

Tamagawa Numbers

In this lecture and the next, we summarize the arguments given in [**Wei61**] to prove that the Tamagawa number $\tau(H)$ of the algebraic group H associated to the group of elements of reduced norm 1 in a division algebra D over a number field k is 1. In keeping with the principle of saving time by focusing on the main ideas, we will discuss only the proof of

THEOREM 1.0.1. *Let D^* denote the group of invertible elements in D, and let $Z^* \subset D^*$ be the center. We view D^* and Z^* as algebraic groups over k, and define G to be the quotient algebraic group$G = D^*/Z^*$. Then the Tamagawa number $\tau(G) = n$, where $n^2 = [D{:}k]$.*

The isogeny argument showing $\tau(G) = n \iff \tau(H) = 1$ is given on [**Wei61**, pp. 48–50]. It will be omitted here.

1.1. The Tamagawa measure.

EXAMPLE 1.1.1. Let k be a number field and let $\mathbb{A}_k$ denote the adeles of k. Define Haar measures dx_ν on the completions k_ν by

$$dx_\nu = dx \qquad k_\nu = \mathbb{R}$$

$$dx_\nu = i dz \wedge d\bar{z} \qquad k_\nu = \mathbb{C}$$

$$\int_{\mathcal{O}_\nu} dx_\nu = 1 \qquad k_\nu \text{ non-Archimedean, } \mathcal{O}_\nu \subset k_\nu \text{ ring of integers.}$$

For S a finite set of valuations containing the Archimedean ones, $\prod_\nu dx_\nu$ defines a Haar measure on $\prod_{\nu\in S} k_\nu \times \prod_{\nu\in S} \mathcal{O}_\nu$, and hence (taking the limit over larger and larger S) a Haar measure w_A on $\mathbb{A}_k$. Recall $k \subset \mathbb{A}_k$ is discrete.

PROPOSITION 1.1.2. *$\mathbb{A}_k/k$ is compact, and $\mu_k = \int_{\mathbb{A}_k/k} \omega_A = |D_k|^{1/2}$ where $D_k =$* discrimimant *$k/\mathbb{Q}$.*

PROOF. Let $S =$ infinite places of k, $\mathbb{A}_S = \prod_{\nu\in S} k_\nu \times \prod_{\nu\notin S} \mathcal{O}_\nu$. By the approximation theorem, $\mathbb{A}_k = \mathbb{A}_S + k$, so

$$\mathbb{A}_k/k \cong \mathbb{A}_S/k \cap \mathbb{A}_S \cong (k \otimes_{\mathbb{Q}} \mathbb{R}/\mathcal{O}_k) \times \prod_{\nu\notin S} \mathcal{O}_\nu$$

where $\mathcal{O}_k \subset k$ is the ring of integers.

This proves compactness, and shows

$$\mu_k = \int_{k\otimes_{\mathbb{Q}}\mathbb{R}/\mathcal{O}_k} \prod_{\nu\in S} dx_\nu .$$

If $\{a_\alpha\}$ is a basis for $\mathcal{O}_k$ over $\mathbb{Z}$ and $\sigma_\beta\colon k \hookrightarrow \mathbb{R}$ (resp. $\sigma_\gamma\colon k \hookrightarrow \mathbb{C}$) are the distinct conjugacy classes of real (resp. complex) embeddings, $1 \le \beta \le r_1$, $1 \le \gamma \le r_2$, then one sets

$$\mu_k = \left|\det\bigl(\sigma_\beta(a_\alpha), \sigma_\gamma(a_\alpha), \overline{\sigma_\gamma(a_\alpha)}\bigr)\right|.$$

Let M denote the matrix in parentheses. Then

$$M^{\mathrm{t}}M = \bigl(\mathrm{tr}_{k/\mathbb{Q}}(a_\alpha a_{\alpha'})\bigr)$$

so $\mu_k^2 = |\Delta_k|$. □

1.2. More generally, let V be an algebraic variety of dimension n over k, $x^0 \in V$ a smooth point, ν a valuation on k, and ω a differential n-form on V. Assume x^0 and ω are defined over k, and $\omega(x^0) \neq 0$. We choose local coordinates $x_1, \ldots, x_n$ and define an isomorphism from a neighborhood of the origin in k_ν^n to a (ν-adic) neighborhood of x^0 on $V(k_\nu)$

$$x \mapsto (x_1 - x_1^0, \ldots, x_n - x_n^0).$$

We could use this isomorphism to define a measure $dx_{1_\nu} \cdots dx_{n_\nu}$ near x on V, but this would depend upon the choice of local coordinates. The clever thing to do is to write

$$\omega = f\, dx_1 \wedge \cdots \wedge dx_n$$

near x^0 and take the measure to be

$$|f|_\nu\, dx_{1_\nu} \cdots dx_{n_\nu}.$$

To check that *this* measure is independent of local coordinates, we can change coordinates one at a time and so reduce to considering a new set $y_1, x_2, \ldots, x_n$. We can then integrate à la Fubini by fixing values for $x_2, \ldots, x_n$ and integrating first with respect to x_1 (resp. y_1), thereby reducing to the case $n = 1$. Since for $a \in k_\nu$

$$f\, dx = a^{-1} f\, d(ax)$$

we may assume $y = x + a_2x^2 + \cdots$, $x^0 = 0$. In this case

$$f\, dx = f\frac{dx}{dy}dy,$$

$dy = dx$, and $|dx/dy| = 1$ near 0, so we are done.

1.3. Let V be smooth of dimension n over a number field k, and let ω be a non-vanishing n-form on V also defined over k. We can find a finite set S of (bad) primes of k and a model $V_{\mathcal{O}_{k,S}}$ of V over the ring $\mathcal{O}_{k,S}$ of S-integers, together with a non-vanishing n-form ω_S on $V_{\mathcal{O}_{k,S}}$. If $p \notin S$, and $\mathbb{F}_q = \mathcal{O}_p/p\mathcal{O}_p$, we can consider the sets $V(\mathcal{O}_p)$ and $V(\mathbb{F}_q)$ of points of $V_{\mathcal{O}_{k,S}}$ with values in $\mathcal{O}_p$ and $\mathbb{F}_q$ respectively. We view $V(\mathcal{O}_p)$ as a p-adic topological space. As such, it is a disjoint union of balls, and isomorphic to

$$\underbrace{(p) \times \cdots \times (p)}_{n \text{ times}},$$

one for each point of $V(\mathbb{F}_q)$. As above, ω induces a measure ω_p on $V(\mathcal{O}_p)$. With notation as above, non-vanishing of ω means $|f| = 1$ on the neighborhood where the x_i's are local coordinates. By definition $\int_{\mathcal{O}_p} dx = 1$, so $\int_{(p)} dx = q^{-1}$, and

$$\int_{V(\mathcal{O}_p)} \omega_p = q^{-n} \cdot \#V(\mathbb{F}_q). \tag{1.3.1}$$

1.4. The Global Measure and Convergence Factors. A *set of factors* is a collection of positive real numbers $\{\lambda_\nu\}$, one for each place of k. These are a *set of convergence factors* if

$$\prod_p \left(\lambda_p^{-1} \int_{V(\mathcal{O}_p)} \omega_p \right)$$

converges absolutely. One checks that this notion is independent of the model $V_{\mathcal{O}_{k,S}}$ chosen for V_k, and is independent of the choice of non-vanishing n-form ω.

Let $V(\mathbb{A}_k)$ denote the adele-valued points of V (i.e. $x \in V(\mathbb{A}_k)$ is a collection $x = (x_\nu)$, $x_\nu \in V(k_\nu)$, $x_p \in V(\mathcal{O}_p)$ for almost all finite primes p. Note that changing the model $V_{\mathcal{O}_{k,S}}$ changes $V(\mathcal{O}_p)$ only for a finite number of p, so $V(\mathbb{A}_k)$ is independent of the choice of model.)

DEFINITION 1.4.1. Given a set of convergence factors $\{\lambda_\nu\}$ and a non-vanishing n-form ω on V_k, the *Tamagawa measure* $(\omega, \{\lambda_\nu\})$ on $V(\mathbb{A}_k)$ is defined by passing to the limit over larger and larger finite sets of places S of the product measure $\mu_k^{-n} \prod_\nu (\lambda_\nu^{-1} \omega_\nu)$ on

$$\prod_{\nu \in S} V(k_\nu) \times \prod_{\nu \notin S} V(\mathcal{O}_\nu).$$

Here μ_k is the volume of $\mathbb{A}_k/k$ as defined in Proposition 1.1.2.

It follows from the product formula that $(\omega, \{\lambda_\nu\}) = (c\omega, \{\lambda_\nu\})$ for $c \in k^*$.

1.5. Restriction of Scalars. To understand Borel's paper, one must understand the role of μ_k in the formula for the Tamagawa measure, and this involves *restriction of scalars.* Let K/k be a finite extension field. Given V defined over K, the restriction of scalars $R_{K/k}V$ is a variety over k together with a canonical identification on T-points for any k-scheme T

$$R_{K/k}V(T) = V(T \underset{k}{\times} K).$$

In particular

$$(R_{K/k}V) \underset{k}{\times} \bar{k} = \prod_{\sigma} V^{\sigma}$$

where σ runs through embeddings $K \to \bar{k}$ over k, and V^{σ} denotes the $\bar{k}$-variety defined via σ. In particular if $n = \dim_K V$ and $d = [K{:}k]$, $\dim_k R_{K/k}V = nd$.

The functor of points property implies

$$V(\mathbb{A}_K) = (R_{K/k}V)(\mathbb{A}_k) \qquad \text{(exercise!)} \tag{1.5.1}$$

Given convergence factors $\{\lambda_w\}$ for V over K, and a non-vanishing n-form ω, it is natural to look for a non-vanishing nd-form $R_{K/k}\omega$ on $R_{K/k}V$, and to expect $\lambda_\nu = \prod_{w|\nu} \lambda_w$ to be a set of convergence factors for $R_{K/k}\omega$. The problem is that the natural differential to choose, $\bigwedge_\sigma \omega^\sigma$ on $(R_{K/k}V)_{\bar{k}}$ depends on the ordering of the set of σ's. To remove this dependence, we write $K = k(\theta)$, $\sqrt{\Delta} = \prod_{i \leq j}(\theta^{\sigma_i} - \theta^{\sigma_j})$ for some ordering of the σ's, and

$$R_{K/k}\omega = \left(\sqrt{\Delta}\right)^n \bigwedge_{i=1}^{d} \omega^{\sigma_i}.$$

This is well-defined up to a scalar in k^*, and is defined over k. Moreover, the Tamagawa measure $(R_{K/k}\omega, \{\lambda_v\})$ is defined, and under the identification of (1.5.1) we have $(\omega, \{\lambda_w\}) = (R_{K/k}\omega, \{\lambda_v\})$.

1.6. Algebraic Groups and Tamagawa Numbers. When $V = G$ is an algebraic group over k, there is a non-vanishing left-invariant n-form ω unique up to scalars. For some G (e.g. $G =$ special or projective linear group associated to a division algebra D over k) it will turn out that $\lambda_\nu = 1$ is a set of convergence factors for G. $\big(\omega, (1)\big)$ will be called *the* Tamagawa measure of $G(\mathbb{A}_k)$, and the Tamagawa number $\tau(G)$ is defined by

$$\tau(G) = \int_{G(\mathbb{A}_k)/G(k)} \big(\omega, (1)\big).$$

Notice, in general, if $\{\lambda_v\}$ is a set of convergence factors for an algebraic group G over k and ω is a left-invariant differential, then $(\omega, (\lambda_v))$ is a left Haar measure on $G(\mathbb{A}_k)$. Also, right translation by $g \in G(k)$ changes ω to $\psi(g)\omega$, $\psi(g) \in k^*$, so the Tamagawa measure does not change (product rule) and $\int_{G(\mathbb{A}_k)/G(k)}$ makes sense.

LECTURE 2

Tamagawa Numbers. Continued

2.1. Let D be a central division algebra over a number field k, with $[D{:}k] = n^2$. Let D^* denote the algebraic group over k of invertible elements of D. Let $Z^* \subset D^*$ be the center, so $Z^* \cong \mathbb{G}_{m,k}$, and let $G = D^*/Z^*$. For convenience, we will write $I_k = \mathbb{G}_m(\mathbb{A}_k)$, the idèle group. We want to prove $\tau(G) = n$. We will sketch the argument first, postponing proofs.

PROPOSITION 2.1.1. *A set of convergence factors for D^* is provided by*

$$\lambda_v = \begin{cases} 1 & v \text{ infinite} \\ 1 - N(v)^{-1} & v \text{ finite, } N(v) = \#(\text{residue field } \mathcal{O}_v). \end{cases}$$

Note that $\{\lambda_v\}$ does not depend on D. Since $Z = \text{Center}(D)$ is also a division algebra, we get measures ω_{D^*} and ω_{Z^*} with the same convergence factors. This implies that $\lambda_v = 1$ is a system of convergence factors on $G = D^*/Z^*$. We write ω_G for the Tamagawa measures.

Consider now the maps

$$D^*(\mathbb{A}_k) \xrightarrow{N} I_k \xrightarrow{|\,|} \mathbb{R}_t^*$$

where N is the *reduced norm* of the division algebra, and $|\,|$ is the idèle module (normalized in the usual way, so $|k^*| = 1$). We can, if we like, view $|N|$ as a map $D^*(\mathbb{A}_k)/D^*(k) \to \mathbb{R}_+^*$.

PROPOSITION 2.1.2. *Given $0 < m \le M$, the set of points x of $D^*(\mathbb{A}_k)/D^*(k)$ determined by $m \le |N(x)| \le M$ is compact.*

The key fact we need is

THEOREM 2.1.3. *Let F be a function on $\mathbb{R}_+$ such that the integral $\int_0^\infty F(t)\,dt/t$ converges absolutely. Then*

$$\int_{D^*(\mathbb{A}_k)/D^*(k)} F(|N(x)|)\omega_{D^*} = \rho_k \int_0^\infty F(t)\frac{dt}{t},$$

where $\rho_k \neq 0$ and ρ_k depends only on k (and not on D or n).

Granting, for a moment, these results, the Tamagawa number computation is easy. In fact, the exact sequence

$$0 \to Z^* \to D^* \to G \to 1$$

admits local sections (because $Z^* \cong \mathbb{G}_m$) so we get

$$0 \to Z^*(\mathbb{A}_k) \to D^*(\mathbb{A}_k) \to G(\mathbb{A}_k) \to 1$$
$$0 \to Z^*(k) \to D^*(k) \to G(k) \to 1.$$

Given a function F on $\mathbb{R}_+$ as above, we can thus write

$$\int_{D^*(A)/D^*(k)} F(|N(x)|)\omega_{D^*} = \int_{G(A)/G(k)} \omega_G \int_{Z^*(A)/Z^*(k)} F(|N(yz)|)\omega_{Z^*}.$$

(We are using here the fact that the convergence factors chosen for total space D^*, base G, and fibre Z^* are compatible.) By Theorem 2.1.3

$$\int_{D^*(A)/D^*(k)} F(|N(x)|)\omega_{D^*} = \rho_k \int_0^\infty F(t)\frac{dt}{t}$$

$$\int_{Z^*(A)/Z^*(k)} F(|N(yz)|)\omega_{Z^*} = \int_{Z^*(A)/Z^*(k)} F(|N(y)|\,|z|^n)\omega_{Z^*}$$
$$= \rho_k \int_0^\infty F(|N(y)|t^n)\frac{dt}{t} = \frac{\rho_k}{n}\int_0^\infty F(t)\frac{dt}{t}$$

Comparing these values, we find

$$\int_{G(A)/G(k)} \omega_G = n = \tau(G).$$

This completes the proof of Theorem 1.0.1.

2.2. Some Proofs.

PROOF OF PROPOSITION 2.1.1. We can find a finite set of places S of k and a model $D^*_{\mathcal{O}_{k,S}}$ of D^* such that $D^*_{\mathcal{O}_p} \cong \mathrm{GL}_{n,\mathcal{O}_p}$ for all $p \notin S$. By (1.3.1), it will suffice to show that the product

$$\prod_{p \notin S} (1 - N(p)^{-1})^{-1} \cdot N(p)^{-n^2} \cdot \#\,\mathrm{GL}_n\big(k(p)\big) \tag{2.2.1}$$

converges absolutely, where $k(p)$ is the residue field of $\mathcal{O}_p$. For convenience, write $q = N(p)$.

$$\#\,\mathrm{GL}_n\big(k(p)\big) = (q^n - 1)(q^n - q)\cdot \cdots \cdot (q^n - q^{n-1})$$
$$= q^{n^2}(1 - q^{-1}) \cdot \cdots \cdot (1 - q^{-n}).$$

Convergence of (2.2.1) follows from convergence of the zeta function of k, $\zeta_k(s)$, for $\operatorname{Re} s > 1$. □

PROOF OF PROPOSITION 2.1.2. Let D (without the $*$) denote the additive group of the division algebra. The map $x \mapsto (x, x^{-1})$ identifies $D^*(A)$ as a closed subspace of $D(A) \times D(A)$. Let $C \subset D(A)$ have measure $> \max(M^n, m^{-n})$, $C' = C + (-C) = \{c_1 - c_2 \mid c_1, c_2 \in C\}$. For $a \in D^*(A)$, the automorphisms $x \mapsto ax$, $x \mapsto xa$ of $D(A)$ have modules $|N(a)|^n$ ($N(a)$ is the *reduced* norm; its nth power is the norm in the algebra D). Since the measure of $D(A)/D(k)$ is 1 (because of the factor μ_k^{-n} in Definition 1.4.1) we find for $m \le |N(a)| \le M$ the sets $a^{-1}C$, Ca do not map one-to-one to $D(A)/D(k)$, i.e., there exist $\alpha, \beta \in D^*(k)$ with $c' = a\alpha$ and $c'' = \beta a^{-1}$ in C'. But $\beta\alpha = c''c' \in C'^2 \cap D^*(k) =$ finite set, since $D(k)$ is discrete and C'^2 is compact. Hence

$$(a\alpha, \alpha^{-1}a^{-1}) = (c', (\beta\alpha)^{-1} \cdot c'') \in (C' \times (\text{finite set}) \cdot C') \cap D^*(A) = S.$$

The set S on the right is compact, so we have a compact set S in $D^*(A)$ such that for all $a \in D^*(A)$, $m \le |N(a)| \le M$, there exists $\alpha \in D^*(k)$ with $a\alpha \in S$. □

2.3. Start of Proof of Theorem 2.1.3—The Zeta Function. With notation as in Theorem 2.1.3, ω_{D^*} and dt/t are Haar measures on $D^*(A)$ and $\mathbb{R}_+^*$, and $|N|$ is a homomorphism, so it follows from Proposition 2.1.2 that

$$\int_{D^*(A)/D^*(k)} F(|N(x)|)\omega_{D^*} = c\int_0^\infty F(t)\frac{dt}{t} \tag{2.3.1}$$

for some c independent of F. Via a zeta function argument we will exhibit an F for which both sides of (2.3.1) can be calculated.

On the real vector space $D(k \otimes_{\mathbb{Q}} \mathbb{R})$ we fix a positive definite quadratic form f and a polynomial function P, and define

$$\Phi_0 \colon D(k \otimes_{\mathbb{Q}} \mathbb{R}) \to \mathbb{C}$$
$$\Phi_0(x) = P(x)e^{-f(x)}.$$

Choose a model for D over $\mathcal{O}_{k,S}$ for some finite set S of finite primes, and take

$$\Phi_S \colon D\Big(\prod_{p \in S} k_p\Big) \to \mathbb{C}$$

to be any locally constant function with compact support. For $p \notin S$, define Φ_p to be the characteristic function of $D(\mathcal{O}_p)$. Then

$$\Phi = \Phi_0 \cdot \Phi_S \cdot \prod_{p \notin S} \Phi_p$$

is a function on $D(A)$.

DEFINITION 2.3.1. The zeta function of D with respect to the chosen function Φ, $Z^\Phi(s)$, is defined by

$$Z^\Phi(s) = \int_{D^*(A)} |N(x)|^s \Phi(x) \omega_{D^*}.$$

PROPOSITION 2.3.2. *$Z^\Phi(s)$ is absolutely convergent for* $\operatorname{Re} s > n$, *and*

$$\lim_{s \to n^+} Z^\Phi(s) = \rho_k \int_{D(A)} \Phi(x)\, dx_A,$$

where dx_A denotes the Tamagawa measure on $D(A)$ (with respect to which the measure of $D(A)/D(k) = 1$), and ρ_k is a non-zero scalar independent of D and n.

PROOF. Change notation a bit and take for S a finite set including the infinite primes such that $D_{\mathcal{O}_p} \cong M_{n,\mathcal{O}_p}$ ($n \times n$ matrices) and $\Phi_p =$ characteristic function of $D(\mathcal{O}_p)$, for $p \notin S$. With this notation, $\Phi = \Phi_S \cdot \prod_{p \notin S} \Phi_p$.

We now fix a translation-invariant differential ω on D^*

$$\omega = N(x)^{-n}\, dx_{11} \wedge dx_{12} \wedge \cdots \wedge dx_{nn}$$

for some basis x_{ij} of D over $\mathcal{O}_{k,S}$. We write

$$dx_S = \mu_k^{-n^2}\, dx_{11} \wedge dx_{12} \wedge \cdots \wedge dx_{nn},$$

viewed as a measure on $D(\prod_{\nu \in S} k_\nu) = D_S$. For $p \notin S$, let ω_p denote the measure on $D^*(k_p)$ defined by $(1 - N(p)^{-1})^{-1}\omega$. Define for $\operatorname{Re} s > n$

$$\begin{aligned} Z_S^\Phi(s) &= \int_{D_S^*} \prod_{v \in S} |N(x_v)|_v^{s-n} \Phi_S(x_S)\, dx_S \\ &= \text{the same integral taken over } D_S. \end{aligned}$$

$$Z_p(s) = \int_{D^*(k_p)} |N(x_p)|_p^s \Phi_p(x_p) \omega_p,$$

so

$$Z^\Phi(s) = \left(\prod_{v \in S} (1 - N(v)^{-1})^{-1} \right) Z_S^\Phi(s) \prod_{p \notin S} Z_p(s). \tag{2.3.2}$$

Since $\int_{\mathcal{O}_p} dx_p = 1$, we find

$$Z_S^\Phi(n) = \int_{D_S} \Phi_S(x_S)\, dx_S = \int_{D(\mathbb{A}_k)} \Phi(x)\, dx_{\mathbb{A}_k}. \tag{2.3.3}$$

On the other hand, since Φ_p is the characteristic function of $D(\mathcal{O}_p) = M_n(\mathcal{O}_p)$, we find

$$Z_p(s) = \int_{M_n(\mathcal{O}_p)^*} (1-q^{-1})^{-1} |\det X|_p^{s-n} (dX)_p$$

where $M_n(\mathcal{O}_p)^*$ denotes the set of matrices with non-zero determinant, and $q = N(p)$.

To compute $Z_p(s)$, write $U = \mathrm{GL}_n(\mathcal{O}_p)$ and decompose into cosets $M_n(\mathcal{O}_p) = \bigcup UA$ where A denotes an element of the set of matrices

$$\begin{pmatrix} \pi^{d_1} & & & \alpha_{ij} \\ & \ddots & & \\ & & \ddots & \\ 0 & & & \ddots \\ & & & & \pi^{d_n} \end{pmatrix}$$

with $\pi \in \mathcal{O}_p$ prime, $d_1, \ldots, d_n$ running through all n-tuples of integers ≥ 0 and α_{ij} running through a complete set of representatives for $\mathcal{O}_p \bmod \pi^{d_j}$. We find (using the invariance of the measure)

$$\begin{aligned}
(2.3.4)\quad Z_p(s) &= (1-q^{-1})^{-1} \sum_A |\det A|_p^s \int_{UA} |\det X|_p^{-n} (dX)_p \\
&= (1-q^{-1})^{-1} \sum_A |\det A|_p^s \int_U |\det X|_p^{-n} (dX)_p \\
&= (1-q^{-1})^{-1} \sum_A |\det A|_p^s \int_U (dX)_p \\
&= (1-q^{-1})^{-1} \sum_{(d_i)} q^{-sd_1+(-s+1)d_2+\cdots+(-s+n-1)d_n} \\
&\qquad\qquad \times (1-q^{-1}) \cdots (1-q^{-n}) \\
&= (1-q^{-1})^{-1} \frac{(1-q^{-1}) \cdots (1-q^{-n})}{(1-q^{-s})(1-q^{-s+1}) \cdots (1-q^{-s+n-1})}
\end{aligned}$$

so $Z_p(n) = (1-q^{-1})^{-1}$. Using $\lim_{s\to 1}(s-1)\zeta_k(s) = \rho_k \neq 0, \infty$, together with (2.3.2), (2.3.3) and (2.3.4), we easily deduce Proposition 2.3.2. □

We now write $A = \mathbb{A}_k$, and fix a character $\chi\colon A \to$ Circle such that $x \to \chi(x\cdot)$ identifies A with its Pontryagin dual, and such that k is its own perpendicular under this pairing. For the construction of such a χ, see [**Wei67**, p. 66]. Let $\mathrm{Tr}\colon D(A) \to A$ denote the reduced trace, and define

$$\chi_D\colon D(A) \to \text{Circle},$$
$$\chi_D(x) = \chi\big(\mathrm{Tr}(x)\big).$$

Then χ_D identifies $D(A)$ with its Pontryagin dual and $D(k) = D(k)^{\perp}$. The Fourier transform ψ of Φ is defined by

$$\psi(y) = \int_{D(A)} \Phi(x)\chi_D(xy)\,dx_A. \tag{2.3.5}$$

One shows that there exists a finite set S' of places such that $\psi(y) = \psi_{S'}(y_{S'}) \cdot \prod_{p\notin S'} \psi_p(y_p)$, where ψ_p = characteristic function of $D(\mathcal{O}_p)$, and $\psi_{S'}$ is described analogously to Φ_S above. In particular, the zeta function $Z^{\psi}(s)$ is defined.

For $a \in D^*(A)$, $d(ax)_A = |N(a)|^n\,dx_A$, and $\chi_D(axa^{-1}) = \chi_D(x)$. Substituting in (2.3.5), it follows that the Fourier transform of $x \mapsto \Phi(ax)$ is $|N(a)|^{-n}\psi(xa^{-1})$. Also we have the Poisson summation formula

$$\sum_{x\in D(k)} \Phi(x) = \sum_{x\in D(k)} \psi(x). \tag{2.3.6}$$

Sketch of proof of (2.3.6). If G is a locally compact group, G^* the dual group, f a "nice" function on G, $\hat{f}$ its Fourier transform on G^*, then $\hat{f}(0) = \int_G f\,d\mu$. Apply this to $G = D(k)$ (discrete), $G^* = D(A)/D(k)$, $f = \Phi$. Note that the measure of $G^* = D(A)/D(k)$ is 1, so the dual measure on $D(k)$ gives every point mass 1. Thus $\hat{f}(0) = \sum_{x\in D(k)} \Phi(x)$. On the other hand, the Fourier transform of the function F on G^* defined by $F(\bar{a}) = \sum_{x\in D(k)} \psi(x+a)$ is given by

$$\begin{aligned}\widehat{F}(y) = \int_{G^*} F(u)\langle u,y\rangle\,d\mu &= \int_{D(A)/D(k)} \langle u,y\rangle\,d\mu \sum_{x\in D(k)} \psi(x+u) \\ &= \int_{D(A)} \psi(u)\langle u,y\rangle\,dx_A = \Phi(-y).\end{aligned}$$

Hence $\widehat{F}(y) = f(-y)$, $F = \hat{f}$, and $\hat{f}(0) = \sum_{x\in D(k)} \psi(x)$.) □

Define a function f on $\mathbb{R}_+$ by

$$f(t) = \begin{cases} 1 & \text{if } t < 1 \\ \frac{1}{2} & \text{if } t = 1 \\ 0 & \text{if } t > 1. \end{cases}$$

We have $f(t) + f(t^{-1}) \equiv 1$. Define $f_\pm \colon D^*(A) \to \mathbb{R}$ by

$$f_+(x) = f(|N(x)|)$$
$$f_-(x) = f(|N(x)|^{-1}),$$

and set

$$Z_\pm^\Phi(s) = \int_{D(A)^*} f_\pm(x)|N(x)|^s \Phi(x)\omega_{D^*},$$

so $Z^\Phi = Z_+^\Phi + Z_-^\Phi$. Note the integral in Z_+^Φ converges absolutely and uniformly for all s. We can use Poisson summation to write

$$\begin{aligned}
Z_-^\Phi(s) &= \int_{D^*(A)/D(k)^*} f_-(x)|N(x)|^s \Big(\sum_{a\in D^*(k)} \Phi(xa) \Big) \omega_{D^*} \\
&= \int f_- |N|^s \Big(\sum_{a\in D(k)} \Phi(xa) - \Phi(0) \Big) \omega \\
&= \int f_- |N|^s \Big(|N(x)|^{-n} \sum_{b\in D(k)} \psi(bx^{-1}) - \Phi(0) \Big) \omega \\
&= \int f_-(x)|N(x)|^{s-n} \Big(\sum_{b\in D^*(k)} \psi(bx^{-1}) \Big) \omega \\
&\qquad - \int f_-(x)|N(x)|^s \big(|N(x)|^{-n}\psi(0) - \Phi(0) \big) \omega \\
&= \int_{D^*(A)} f_+(y)|N(y)|^{n-s}\psi(y)\omega \\
&\qquad - c\int_{\mathbb{R}_+} f(t^{-1}) \big(t^{s-n}\psi(0) - t^s\Phi(0) \big) \frac{dt}{t}
\end{aligned}$$

where c is the same constant as in (2.3.1). This gives

$$Z_-^\Phi(s) - Z_+^\psi(n-s) = c\Big(\frac{\psi(0)}{s-n} - \frac{\Phi(0)}{s} \Big).$$

From Proposition 2.3.2 we get

$$\begin{aligned}
\rho_k \int \Phi(x) dx_A = \rho_k \psi(0) &= \lim_{s\to n} Z^\Phi(s) \\
&= \lim_{s\to n} (s-n)\big(Z_-^\Phi(s) + Z_+^\Phi(s) \big) \\
&= \lim_{s\to n} (s-n)\big(Z_-^\Phi(s) \big) \\
&= \lim_{s\to n} (s-n)\big(Z_-^\Phi(s) - Z_+^\psi(n-s) \big) = c\psi(0).
\end{aligned}$$

Assuming Φ chosen so $\psi(0) \neq 0$ (easy), we get $c = \rho_k$. In particular c is independent of n and D. This completes the proof of Theorem 1.0.1.

LECTURE 3

Continuous Cohomology

3.1. For G a Lie group and A a continuous G-module, it is natural to conceive of a cohomology theory $H^*(G, A)$ analogous to the Eilenberg-MacLane theory for discrete groups but taking into account the topology of G. It turns out that there is a nice theory, which we will denote $H^*_{\text{cont}}(G, A)$ and refer to as the *continuous cohomology* of G. In fact (under reasonable hypotheses) the groups $H^*_{\text{cont}}(G, A)$ can be computed using differentiable, continuous, or even measurable cochains. For example, we will see in Lecture 7 that the imaginary part of the dilogarithm of the cross-ratio defines a measurable function

$$D\colon \underbrace{\mathrm{SL}_2(\mathbb{C}) \times \cdots \times \mathrm{SL}_2(\mathbb{C})}_{4 \text{ times}} \to \mathbb{R} \tag{3.1.1}$$

satisfying the homogeneous cocycle identity

$$\sum (-1)^i D(x_0, \ldots, \hat{x}_i, \ldots, x_5) = 0, \tag{3.1.2}$$

and hence a class in $H^3_{\text{cont}}(\mathrm{SL}_2(\mathbb{C}), \mathbb{R})$. An even simpler example: the real part of the logarithm defines a class in $H^1_{\text{cont}}(\mathrm{GL}_1(\mathbb{C}), \mathbb{R})$.

3.2. Let G be a real Lie group, $K \subset G$ a closed compact subgroup. Let $C^q(G/K, \mathbb{R})$ denote the space of C^∞-differential q-forms on G/K. We topologize C^q by $f_i\, dx_1 \wedge \cdots \wedge dx_q \to f\, dx_1 \wedge \cdots \wedge dx_q$ if and only if $f_i \to f$ in the C^∞-topology. G acts on G/K by left translation, and hence on C^q by $(g\omega)(x) = \omega(g^{-1}x)$, and it turns out that C^q is injective in a suitable category of continuous G-modules. We can consider the continuous hypercohomology of G acting on the de Rham complex $C(G/K, \mathbb{R})$. As usual there are two spectral sequences, denoted $'E$ and $''E$, abutting to $H^*_{\text{cont}}\big(G, C^\bullet(G/K, \mathbb{R})\big)$:

$$\begin{aligned} {}'E_1^{p,q} &= H^q_{\text{cont}}\big(G, C^p(G/K, \mathbb{R})\big) \\ {}''E_2^{p,q} &= H^p_{\text{cont}}\left(G, H^q\big(C^\bullet(G/K, \mathbb{R})\big)\right) \end{aligned} \tag{3.2.1}$$

Thus $'E_1^{p,q} = (0)$, $p > 0$, and

$$'E_1^{0,q} = H^q(C^\bullet(G/K, \mathbb{R})^G) \overset{\text{dfn}}{=} H^q(\mathfrak{g}, k; \mathbb{R})$$

where $\mathfrak{g}$, k are the Lie algebras of G and K and the relative Lie algebra cohomology $H^q(\mathfrak{g}, k; \mathbb{R})$ can be defined to be the cohomology of the complex of left-invariant differential forms on G/K. On the other hand, by de Rham $H^q(C^\bullet(G/K, \mathbb{R})) \cong H^q_{\text{top}}(G/K, \mathbb{R})$, the cohomology of the topological space G/K. Thus

$$''E_2^{p,q} = H^p_{\text{cont.}}(G, H^q_{\text{top}}(G/K, \mathbb{R})) \implies H^{p+q}(\mathfrak{g}, k; \mathbb{R}). \tag{3.2.2}$$

3.3. Several Special Cases of Interest.

Case I. *G compact, $K = G$.* We conclude

$$H^p_{\text{cont}}(G, \mathbb{R}) = (0), \quad p > 0. \tag{3.3.1}$$

Case II. *G compact, connected, K arbitrary.* Using Case I, we conclude

$$H^q_{\text{top}}(G/K, \mathbb{R}) \cong H^q(\mathfrak{g}, k; \mathbb{R}). \tag{3.3.2}$$

Case III. *G arbitrary connected, $K \subset G$ maximal compact.* Then $G/K \cong$ Euclidean space, so

$$H^p_{\text{cont}}(G, \mathbb{R}) \cong H^p(\mathfrak{g}, k; \mathbb{R}). \tag{3.3.3}$$

Here are some simple examples we will use later.

EXAMPLE 3.3.1. $G = V =$ finite dimensional $\mathbb{R}$-vector space. We apply (3.3.3) with $k = (0)$, to get

$$H^*_{\text{cont}}(V, \mathbb{R}) \cong \bigwedge\nolimits^*_{\mathbb{R}}(V^*), \quad V^* = \mathrm{Hom}_{\mathbb{R}}(V, \mathbb{R}).$$

EXAMPLE 3.3.2. $G = T \cong \underbrace{\mathbb{C}^* \times \cdots \times \mathbb{C}^*}_{n \text{ times}}$, a torus. We have an exact sequence

$$0 \to (S^1)^n \to T \to \mathbb{R}^n \to 0$$

and since S^1 is compact, $H^*_{\text{cont}}(T, \mathbb{R}) \cong H^*_{\text{cont}}(\mathbb{R}^n, \mathbb{R}) \cong \bigwedge^*_{\mathbb{R}}(\mathbb{R}_n)^*$.

EXAMPLE 3.3.3. Let $B \subset \mathrm{SL}_2(\mathbb{C})$ be the subgroup of upper triangular matrices. There is an exact sequence

$$0 \to \mathbb{C} \to B \rightleftarrows \mathbb{C}^* \to 1$$

and $H^*_{\text{cont}}(B, \mathbb{R}) \cong H^*_{\text{cont}}(\mathbb{C}^*, \mathbb{R})$. Indeed, there is a spectral sequence

$$H^p_{\text{cont}}(\mathbb{C}^*, H^q_{\text{cont}}(\mathbb{C}, \mathbb{R})) \implies H^{p+q}_{\text{cont}}(B, \mathbb{R})$$

Fix $\alpha \in \mathbb{R}^*$ with $|\alpha| \neq 1$ and consider conjugation by the matrix $\left(\begin{smallmatrix} cc\alpha & 0 \\ 0 & \alpha^{-1} \end{smallmatrix}\right)$. This acts trivially on the continuous cohomology of B, and also acts trivially on $\mathbb{C}^*$ viewed as a subgroup of B. The action on $\mathbb{C}$ is by multiplication by α^2. It follows that the above spectral sequence degenerates as claimed.

EXAMPLE 3.3.4. Take $G = \mathrm{SL}_2(\mathbb{C})$, $K = \mathrm{SU}_2(\mathbb{C})$, G/K = Poincaré upper half space. Since $\dim G/K = 3$, the ratio of two translation-invariant 3-forms is a translation-invariant function, i.e., constant, so $\dim H^3_{\mathrm{cont}}(G, \mathbb{R}) \leq 1$. But the volume form on G/K is G-invariant, and the differential on $C^0(G/K, \mathbb{R})^G$ is zero, so $H^3_{\mathrm{cont}}(G, \mathbb{R}) \cong \mathbb{R}$.

3.4. Now let G be a semi-simple connected linear algebraic group defined over $\mathbb{Q}$, and let $\mathfrak{g}$ = Lie algebra. We can use the existence of compact real forms of $G(\mathbb{C})$, [**Hel62**], to compute $H^*_{\mathrm{cont}}(G, \mathbb{C})$. Namely we can decompose

$$\mathfrak{g}(\mathbb{R}) = \mathfrak{K} \oplus \mathfrak{p}$$

where $\mathfrak{K}$ and $\mathfrak{p}$ are the positive and negative eigenspaces of some Cartan involution, and $\mathfrak{K}$ is the Lie algebra of a maximal compact subgroup $K \subset G(\mathbb{R})$. Then

$$\mathfrak{g}_u = \mathfrak{K} \oplus \sqrt{-1}\,\mathfrak{p} \subset \mathfrak{g}(\mathbb{C})$$

is the Lie algebra of a maximal compact $G_u \subset G(\mathbb{C})$. The two pairs of $\mathbb{R}$-Lie algebras

$$\mathfrak{K} \subset \mathfrak{g}_u \quad \text{and} \quad \mathfrak{K} \subset \mathfrak{g}(\mathbb{R})$$

have isomorphic complexifications (namely $\mathfrak{K} \otimes_{\mathbb{R}} \mathbb{C} \subset \mathfrak{g}(\mathbb{C})$) so (needless to say, the previous discussion of $H^*_{\mathrm{cont}}(G, \mathbb{R})$ is also valid for $H^*_{\mathrm{cont}}(G, \mathbb{C})$)

$$(3.4.1) \qquad \begin{aligned} H^*_{\mathrm{cont}}(G(\mathbb{R}), \mathbb{C}) &\cong H^*(\mathfrak{g}(\mathbb{R}), k; \mathbb{C}) \cong H^*(\mathfrak{g}(\mathbb{C}), k \otimes_{\mathbb{R}} \mathbb{C}; \mathbb{C}) \\ &\cong H^*(\mathfrak{g}_u, k; \mathbb{C}) \underset{(3.3.2)}{\cong} H^*_{\mathrm{top}}(G_u/K, \mathbb{C}). \end{aligned}$$

Similarly, dropping $\mathfrak{K}$ and just using $\mathfrak{g}_u \otimes_{\mathbb{R}} \mathbb{C} \cong \mathfrak{g}(\mathbb{R}) \otimes_{\mathbb{R}} \mathbb{C}$, we get

$$(3.4.2) \qquad H^*(\mathfrak{g}(\mathbb{C}), \mathbb{C}) \cong H^*_{\mathrm{top}}(G_u, \mathbb{C}).$$

To conform with Borel's notation, we now quotient out by compact groups acting on the *left* and define

$$X = K\backslash G(R), \quad X_u = K\backslash G_u.$$

We get a commutative square

$$(3.4.3) \qquad \begin{array}{ccc} H^*_{\mathrm{top}}(G_u, \mathbb{C}) & \underset{(3.4.2)}{\xleftarrow{\alpha^*}} & H^*(\mathfrak{g}(\mathbb{C}), \mathbb{C}) \\ \uparrow{\sigma^*} & & \uparrow{\lambda^*} \\ H^*_{\mathrm{top}}(X_u, \mathbb{C}) & \underset{(3.4.1)}{\xleftarrow{\bar{\alpha}^*}} & H^*(\mathfrak{g}(\mathbb{C}), \mathfrak{K} \otimes \mathbb{C}; \mathbb{C}) \cong H^*_{\mathrm{cont}}(G, \mathbb{C}). \end{array}$$

The maps σ^*, λ^* are the evident ones, α^* is obtained by restricting invariant differential forms on G to G_u, and $\bar{\alpha}^*$ is defined analogously.

EXAMPLE 3.4.1. Let k be a number field of degree d over $\mathbb{Q}$, and let $G_n = R_{k/\mathbb{Q}} \mathrm{SL}_{n,k}$. Thus

$$G_n(\mathbb{R}) = \mathrm{SL}_n(k \otimes_{\mathbb{Q}} \mathbb{R}) = \mathrm{SL}_n(\mathbb{R})^{r_1} \times \mathrm{SL}_n(\mathbb{C})^{r_2}$$
$$G_n(\mathbb{C}) = \mathrm{SL}_n(\mathbb{C})^d.$$

With notation as above, $K_n = \mathrm{SO}_n^{r_1} \times \mathrm{SU}_n^{r_2}$ and $G_{n,u} = \mathrm{SU}_n^d$, so

$$X_n = (\mathrm{SO}_n \setminus \mathrm{SL}_n(\mathbb{R}))^{r_1} \times (\mathrm{SU}_n \setminus \mathrm{SL}_n(\mathbb{C}))^{r_2}$$
$$X_{n,u} = (\mathrm{SO}_n \setminus \mathrm{SU}_n)^{r_1} \times \mathrm{SU}_n^{r_2}.$$

The cohomology of these spaces is known. E.g., $H^*_{\mathrm{top}}(\mathrm{SU}_n, \mathbb{Q})$ is an exterior algebra with one primitive generator in degree $2m+1$ for $1 \le m < n$. Similarly $H^*_{\mathrm{top}}(\mathrm{SO}_n, \mathbb{Q})$ is an exterior algebra with generators in degree $2m+1$ for m odd (at least for $n \to \infty$). If n is odd, $H_*(\mathrm{SO}_n, \mathbb{Q}) \hookrightarrow H_*(\mathrm{SU}_n, \mathbb{Q})$. $H^*(\mathrm{SO}_n \setminus \mathrm{SU}_n, \mathbb{Q})$ is also an exterior algebra, and there is an isomorphism of algebras

$$H^*(\mathrm{SO}_n \setminus \mathrm{SU}_n, \mathbb{Q}) \otimes H^*_{\mathrm{top}}(\mathrm{SO}_n, \mathbb{Q}) \to H^*_{\mathrm{top}}(\mathrm{SU}_n, \mathbb{Q}).$$

In the context of Example 3.4.1, if we write $\mathfrak{g}_n = \mathrm{Lie}(G_n)$, $k_n = \mathrm{Lie}\, K_n$, etc., and take n odd, we obtain a diagram of algebra isomorphisms

$$\begin{array}{ccc} H^*_{\mathrm{top}}(G_{n,u}, \mathbb{C}) & \xleftarrow{\quad \alpha^* \quad} & H^*(\mathfrak{g}_n, \mathbb{C}) \\ \uparrow & & \uparrow \\ H^*_{\mathrm{top}}(K_n, \mathbb{C}) \otimes H^*(X_{n,u}, \mathbb{C}) & \longleftarrow & H^*(\mathfrak{g}_n, k_n; \mathbb{C}) \otimes H^*(k_n, \mathbb{C}). \end{array} \tag{3.4.4}$$

If $\bigoplus_{n \ge 1} A^n = A^*$ is a graded algebra, the *indecomposables* of degree n are defined by $I^n(A^*) \stackrel{\mathrm{dfn}}{=} A^n / (\sum_{i=1}^{n-1} A^i \cdot A^{n-i})$. The above discussion shows

PROPOSITION 3.4.2. *$I^{2m+1}\big(H^*_{\mathrm{cont}}(G_n, \mathbb{C})\big)$ has dimension (for n large relative to m and odd)*

$$d_m = \begin{cases} r_2 & m \text{ odd}, \ge 1 \\ r_1 + r_2 & m \text{ even}, \ge 1. \end{cases}$$

Moreover, passing to the limit as $n \to \infty$, $H^*_{\mathrm{cont}}(G_\infty, \mathbb{C})$ is an exterior algebra with d_m generators in dimension $2m+1$.

3.5. Note

$$H^*(\mathfrak{g}_n, \mathbb{C}) \cong H^*(\mathfrak{g}_n, \mathbb{Q}) \otimes \mathbb{C}$$

and

$$H^*_{\mathrm{top}}(G_{n,u}, \mathbb{C}) \cong H^*_{\mathrm{top}}(G_{n,u}, \mathbb{Q}) \otimes \mathbb{C}.$$

We want to study the behavior of α^* on these $\mathbb{Q}$-structures. To this end, let

$$P^{2m+1}(G_{n,u},\mathbb{Q}) \subset H^{2m+1}_{\text{top}}(G_{n,u},\mathbb{Q})$$
$$P^{2m+1}(\mathfrak{g}_n,\mathbb{Q}) \subset H^{2m+1}(\mathfrak{g}_n,\mathbb{Q})$$

denote the spaces of primitive elements. By the structure theory of these cohomology algebras, $P^{2m+1} \underset{\cong}{\to} I^{2m+1}$ so $\dim P^{2m+1} = d = [k{:}\mathbb{Q}]$ (assuming $m < n$). Define $L_m(\mathfrak{g}_n,\mathbb{Q})$ and $L_m(G_{n,u},\mathbb{Q})$ to be the dth exterior powers of the corresponding primitive spaces,

$$L_m = \bigwedge^d P^{2m+1}.$$

PROPOSITION 3.5.1.

$$\alpha^* L_m(\mathfrak{g}_n,\mathbb{Q}) = (\pi i)^{d(m+1)}|D|^{-(2m+1)/2} L_m(G_{n,u},\mathbb{Q}),$$

where D denotes the discriminant of $k/\mathbb{Q}$.

PROOF. Consider first the multiplicative group $\mathbb{G}_m$, $\mathbb{G}_m(\mathbb{C}) = \mathbb{C}^*$ and $\mathbb{G}_{m,u} = S^1$. Let $\ell = \text{Lie}(\mathbb{G}_m)$. $H^1(\ell,\mathbb{Q})$ is generated by dt/t and $H_1^{\text{top}}(\mathbb{G}_m(\mathbb{C}),\mathbb{Z}) \cong H_1^{\text{top}}(G_{m,u},\mathbb{Z})$ by a loop γ around the origin. Since $\int_\gamma dt/t = 2\pi i$, we have

$$\alpha^* H^1(\ell,\mathbb{Q}) = (\pi i) H^1_{\text{top}}(\mathbb{G}_{m,u},\mathbb{Q}).$$

Now let $G'_n = \text{SL}_{n,\mathbb{Q}}$ and replace $\mathbb{G}_m$ by $T_n \subset G'_n$ a maximal torus. Let $T_{n,u} = T_n(\mathbb{C}) \cap G'_{n,u}$ be the maximal compact subtorus. The fibration

$$\begin{array}{ccc} T_u & \text{---} & G'_{n,u} \\ & & | \\ & & T_u\backslash G'_{n,u} \end{array}$$

gives rise to a spectral sequence, whose 5-term exact sequence of terms of low degree

$$0 \to H^1(T_u\backslash G'_{n,u}) \to H^1_{\text{top}}(G'_{n,u}) \to H^1_{\text{top}}(T_u) \xrightarrow{\tau} H^2(T_U\backslash G'_{n,u}) \\ \to H^2_{\text{top}}(G'_{n,u})$$

gives a transgression isomorphism $(H^1(G'_{n,u}) = H^2(G'_{n,u}) = (0))$

$$H^1_{\text{top}}(T_u,\mathbb{Q}) \underset{\cong}{\xrightarrow{\tau}} H^2(T_u\backslash G'_{n,u},\mathbb{Q}).$$

A similar situation obtains on the level of Lie algebras, so writing $\mathfrak{h} = \mathrm{Lie}(T)$ we get a commutative diagram of isomorphisms

$$\begin{array}{ccc} H^1(\mathfrak{h}, \mathbb{Q}) & \xrightarrow{\alpha^*} & \pi i H^1_{\mathrm{top}}(T_u, \mathbb{Q}) \\ \downarrow{\scriptstyle\tau} & & \downarrow{\scriptstyle\tau} \\ H^2(\mathfrak{g}'_n, h, \mathbb{Q}) & \xrightarrow{\alpha^*} & \pi i H^2(T_u \backslash G'_{n,u}, \mathbb{Q}). \end{array}$$

We will use two facts without proof. Firstly, $H^{2m}(T_u\backslash G'_{n,u}, \mathbb{Q})$ is generated by products of elements in H^2 [**Bor53**], and secondly in the spectral sequence

$$E_2^{p,q} = H^p(T_u\backslash G'_{n,u}, H^q_{\mathrm{top}}(T_u, \mathbb{Q})) \implies H^{p+q}_{\mathrm{top}}(G'_{n,u}, \mathbb{Q})$$

$P^{2m+1}(G'_{n,u}, \mathbb{Q})$ corresponds to a subquotient $E_\infty^{2m,1}$ of

$$H^{2m}(T_u\backslash G'_{n,u}, H^1_{\mathrm{top}}(T_u, \mathbb{Q}))$$

[**Ler51**]. Of course, similar facts are also true for the Lie algebras. Since α^* is compatible with the multiplicative and spectral sequence structures, we find

$$\alpha^* H^{2m}(\mathfrak{g}_n, \mathfrak{h}; H^1(\mathfrak{h}, \mathbb{Q})) \subseteq (\pi i)^{m+1} H^{2m}\big(T_u\backslash G'_{n,u}; H^1(T_u, \mathbb{Q})\big).$$

By the second fact, we have

$$\begin{array}{ccc} P^{2m+1}(\mathfrak{g}'_n, \mathbb{Q}) & \xrightarrow{\alpha^*} & P^{2m+1}(G'_{n,u}, \mathbb{C}) \\ \cap & & \cap \\ E_\infty^{2m,1}(\mathfrak{g}'_n, \mathbb{Q}) & \xrightarrow{\alpha^*} (\pi i)^{m+1} E_\infty^{2m,1}(G'_{n,u}, \mathbb{Q}) \subset & E_\infty^{2m,1}(G'_{n,u}, \mathbb{C}) \end{array}$$

so

$$\begin{aligned} \alpha^* P^{2m+1}(\mathfrak{g}'_n; \mathbb{Q}) &\subset \big((\pi i)^{m+1} E_\infty^{2m,1}(G'_{n,u}, \mathbb{Q}) \cap P^{2m+1}(G'_{n,u}, \mathbb{C})\big) \\ &= (\pi i)^{m+1} P^{2m+1}(G'_{n,u}, \mathbb{Q}). \end{aligned} \tag{3.5.1}$$

Now consider the group $G_n = R_{k/\mathbb{Q}} \mathrm{SL}_{n,k}$. If ω_{2m+1} is a generator of $P^{2m+1}(\mathfrak{g}'_n, \mathbb{Q})$, we have seen that

$$R_{k/\mathbb{Q}}\omega_{2m+1} = |D|^{-(2m+1)/2} \bigwedge_\sigma {}^\sigma\omega_{2m+1} \tag{3.5.2}$$

is a generator for

$$L_m(\mathfrak{g}_n, \mathbb{Q}) = \bigwedge^d P^{2m+1}(\mathfrak{g}_n, \mathbb{Q}).$$

(Here σ runs through all embeddings of $k \hookrightarrow \mathbb{C}$, and ${}^\sigma\omega_{2m+1}$ denotes the obvious element in the Lie algebra cohomology of $G_{n,\mathbb{C}} = \prod_\sigma {}^\sigma \mathrm{SL}_{n,k}$.) The proposition follows from (3.5.1) and (3.5.2). $\square$

LECTURE 4

A Theorem of Borel and its Reformulation

4.1. Let G_n be as before (to fix ideas, we stick to G_n, though the setup is valid more generally). Let $\Gamma_n \subset G_n(\mathbb{Q})$ be some torsion-free arithmetic subgroup. We have a commutative square
(4.1.1)

$$\begin{array}{ccc} H^*(\mathfrak{g}_n, \mathbb{C}) & \xrightarrow{\quad\beta^*\quad} & H^*(G_n(\mathbb{R})/\Gamma_n, \mathbb{C}) \\ \uparrow{\scriptstyle\lambda^*} & & \uparrow{\scriptstyle\sigma'^*} \\ H^*_{\text{cont}}(G_n(\mathbb{R}), \mathbb{C}) \cong H^*(\mathfrak{g}_n, \mathfrak{K}_n; \mathbb{C}) & \xrightarrow{\bar{\beta}^*} & H^*(X_n/\Gamma_n, \mathbb{C}) = H^*(\Gamma_n, \mathbb{C}) \end{array}$$

where β^* comes from interpreting the Lie algebra cohomology as the cohomology of right-invariant forms on G_n. Such a form on $G_n(\mathbb{R})$ descends to a form on $G_n(\mathbb{R})/\Gamma_n$. Similarly, classes in $H^*(\mathfrak{g}_n, \mathfrak{K}_n; \mathbb{C})$ arise from right-invariant forms which are bi-invariant under K_n, hence descend to forms on $X_n/\Gamma_n = K_n \backslash G_n(\mathbb{R})/\Gamma_n$. Viewed as a map

$$\bar{\beta}^* \colon H^*_{\text{cont}}(G_n(\mathbb{R}), \mathbb{C}) \to H^*(\Gamma_n, \mathbb{C}),$$

$\bar{\beta}^*$ simply restricts continuous cohomology classes on $G_n(\mathbb{R})$ to the subgroup Γ_n.

At this point we need a very difficult theorem of Borel [**Bor74**].

THEOREM 4.1.1. *For $m < (n-1)/4$, the map*

$$\bar{\beta}^* \colon H^m_{\text{cont}}(G_n(\mathbb{R}), \mathbb{C}) \to H^m(\Gamma_n, \mathbb{C})$$

is an isomorphism for any arithmetic subgroup of $G_n(\mathbb{Q})$.

We will have nothing to say about the proof of this result.

COROLLARY 4.1.2. *With hypotheses as above,*

$$\bar{\beta}^* \colon H^m(\mathfrak{g}_n, \mathbb{C}) \to H^m(G_n(\mathbb{R})/\Gamma_n, \mathbb{C})$$

is also an isomorphism.

PROOF. If Γ_n is torsion-free, $\Gamma_n \cap K_n = (1)$ and we have a spectral sequence

$$H^p\big(\Gamma_n, H^q_{\text{top}}(K_n, \mathbb{C})\big) \implies H^{p+q}(G_n(\mathbb{R})/\Gamma_n, \mathbb{C}),$$

which will be isomorphic to the spectral sequence

$$H^p\big(\mathfrak{g}_n, k_n; H^q(k_n, \mathbb{C})\big) \implies H^{p+q}(\mathfrak{g}_n, \mathbb{C})$$

for $p < (n-1)/4$. For Γ_n arbitrary we can find $\Gamma' \subset \Gamma_n$ torsion-free, so we get

$$\begin{array}{ccc} H^m(\Gamma_n, \mathbb{C}) & \hookrightarrow & H^m(\Gamma', \mathbb{C}) \\ & \nwarrow^{\bar{\beta}^*} & \wr \| \bar{\beta}'^* \\ & & H^m_{\mathrm{cont}}(G_n(\mathbb{R}), \mathbb{C}), \end{array}$$

whence $\bar{\beta}^*$ is an isomorphism. □

4.2. We would like to study the behavior of the $\mathbb{Q}$-structure on $H^m(\mathfrak{g}_n, \mathbb{C})$ and $H^m(G_n(\mathbb{R})/\Gamma_n, \mathbb{C})$ under β^*. For convenience, let $Y_n = G_n(\mathbb{R})/\Gamma_n$. Recall $L_m(\mathfrak{g}_n, \mathbb{Q})$ denotes the dth exterior power of the subspace of primitive elements in $H^{2m+1}(\mathfrak{g}_n, \mathbb{Q})$. Let $L_m(Y_n, \mathbb{Q})$ denote the dth exterior power of the quotient space of indecomposables of $H^{2m+1}(Y_n, \mathbb{Q})$. For $n > 4d(2m+1)+1$,

$$\beta^* : L_m(\mathfrak{g}_n, \mathbb{Q}) \otimes \mathbb{C} \xrightarrow{\cong} L_m(Y_n, \mathbb{Q}) \otimes \mathbb{C}.$$

THEOREM 4.2.1. *With the above hypotheses,*

$$\beta^* L_m(\mathfrak{g}_n, \mathbb{Q}) = i^{r_2} \zeta_k(m+1) L_m(Y_n, \mathbb{Q}).$$

PROOF. Granted that β^* is an isomorphism, the assertion is independent of which arithmetic Γ_n we choose. We may assume given a family of torsion-free Γ_q, $q \geq n$, such that $\Gamma_q \cap G_r = \Gamma_r$, $r \leq q$. (We view $G_r \subset G_q$ in the usual way.) We may therefore stabilize and take n as large as we please. We take $n > 4d(m+1)^2$, odd, and Γ_n torsion-free.

Suppose $\beta^* L_j(\mathfrak{g}_n, \mathbb{Q}) = s_j L_j(Y_n, \mathbb{Q})$ for some $s_j \in \mathbb{C}$, $1 \leq j \leq m$. Recall from Lecture 3 we have generators $R_{k/\mathbb{Q}} \omega_{2s+1}$ for $L_s(\mathfrak{g}_n, \mathbb{Q})$. Define

$$\eta_j = \bigwedge_{s=1}^{j} R_{k/\mathbb{Q}} \omega_{2s+1}.$$

A simple argument in linear algebra (omitted) shows that

$$\beta^*(\eta_j) \in s_1 \cdots s_j H^{d(j^2+2j)}(Y_n, \mathbb{Q}).$$

If we exhibit compact subvarieties $Z_j \subset Y_n$ of dimension $d(j^2+2j)$ such that

$$0 \neq \int_{Z_j} \beta^*(\eta_j) \in i^{r_2(j^2+2j)} \prod_{s=1}^{j} \zeta_k(s+1) \cdot \mathbb{Q}, \quad 1 \leq j \leq m,$$

then using that $i^{r_2} \in i^{r_2(2j+1)}\mathbb{Q}$, it follows that $s_j \in i^{r_2}\zeta_k(j+1)\mathbb{Q}$ as desired.

Let D be a division algebra of dimension $(j+1)^2$ over k and trivial at the archimedean places. Let $H' \subset D^*$ be the subgroup of elements of reduced norm 1. View H' as an algebraic group over k, and let $H = R_{k/\mathbb{Q}}H'$. We know [**Wei61**] that $H'(\mathbb{A}_k)/H'(k)$ is compact. Let $\mathbb{A}_f \subset \mathbb{A}_k$ denote the finite adeles, and let $U \subset H'(\mathbb{A}_f)$ denote some non-empty open compact subset. By the strong approximation theorem $H'(\mathbb{A}_k) = H'(k \otimes \mathbb{R})UH'(k)$. Let

$$T = \big(U \cdot H'(k \otimes \mathbb{R})\big) \cap H'(k).$$

T is an arithmetic subgroup, and we have a fibration

$$U \to H'(\mathbb{A}_k)/H'(k) \to H'(k \otimes \mathbb{R})/T.$$

This implies that $H(\mathbb{R})/(\text{any arithmetic subgroup})$ is compact.

We embed $H' \hookrightarrow \mathrm{SL}_{n,k}$ and hence $H \hookrightarrow G_n$ by the regular representation of H' on $D \cong k^{(j+1)^2}$, and take the subvariety

$$Z_j = H(\mathbb{R})/(H(\mathbb{R}) \cap \Gamma_n).$$

We have $\dim Z_j = d(j+1)^2 - d = d(j^2+2j)$.

Note that $\beta^*(\eta_j)|_{Z_j}$ is a form of maximal dimension. To see that it is non-trivial, it suffices to look on the level of Lie algebras, i.e., invariant differential forms. Since the derivative in this complex is zero, it suffices to show $\eta_j|_{h_j} \neq 0$ where $h_j = \mathrm{Lie}(H)$. Let $\eta'_j = \bigwedge_{s=1}^{j} \omega_{2s+1}$ on $\mathfrak{g}'_n$. It suffices to show $\eta'_j|_{h'_j} \neq 0$, $h'_j = \mathrm{Lie}\, H'$. We might as well work over $\mathbb{C}$, and $H'(\mathbb{C}) \cong \mathrm{SL}_{j+1}(\mathbb{C})$. The embedding in SL_n is via the regular representation, which is conjugate to $j+1$ copies of the standard representation so we can think of the embedding as

$$\rho\colon \mathrm{SL}_{j+1} \to \mathrm{SL}_n$$

$$\rho(A) = \begin{pmatrix} I & & & \\ & \overbrace{A \quad\quad\quad}^{j+1} & & \\ & & \ddots & \\ & & & A \end{pmatrix}$$

We can factor this embedding

$$\mathrm{SL}_{j+1} \xhookrightarrow[(f_1,\dots,f_{j+1})]{} (\mathrm{SL}_n)^{j+1} \xrightarrow[\text{multiply}]{} \mathrm{SL}_n$$

where $f_i(A) = \begin{pmatrix} I & & \\ & A & \\ & & I \end{pmatrix}$ with A in the ith place. Since the ω_{2s+1} are primitive, we have

$$\text{multiply}^*(\omega_{2s+1}) \underset{\text{cohomologous}}{\sim} \sum_{i=1}^{j+1} \omega_{2s+1}^i$$

so $\rho^*(\omega_{2s+1}) \sim (j+1)\omega_{2s+1}$ on SL_{j+1}. Finally $\rho^*(\eta_j) \sim (j+1)^j \eta_j$ on SL_{j+1} and this is non-trivial.

Since we only wish to calculate $\int_{Z_j} \rho^*(\eta_j)$ up to a rational factor, we are free to replace $\beta^*(\eta_j)|_{Z_j}$ by any volume form defined over $\mathbb{Q}$. In other words, we take on H' some $j^2 + 2j$-form ω' which is non-zero, translation-invariant, and defined over k. We define

$$\omega = |D|^{-(j^2+2j)/2} \bigwedge_{\sigma} {}^{\sigma}\omega'$$

on H. Look at the fibration as above

$$U \to H(\mathbb{A}_{\mathbb{Q}})/H(\mathbb{Q}) \to H(\mathbb{R})/T.$$

Let ω_f and ω_∞ denote the measures on U and $H(\mathbb{R})/T$ induced by ω. Notice that ω_∞ is obtained by integration with respect to $i^{r_2(j^2+2j)}\omega$ on $H(\mathbb{R})/T$, the factor of i arising from the normalization $dx_\nu = i\, dz \wedge d\bar{z}$ in Example 1.1.1. Finally, using (1.3.1) and

$$\#\,\mathrm{SL}_{j+1}(k(p)) = q^{(j+1)^2-1}(1-q^{-2})\cdots(1-q^{-j-1})$$

we find

$$\int_U \omega_f \in \zeta_k(2)^{-1} \cdots \zeta_k(j+1)^{-1} \cdot \mathbb{Q}.$$

Since

$$\int_U \omega_f \cdot \int_{H(\mathbb{R})/T} \omega_\infty = \int_{H(\mathbb{A}_{\mathbb{Q}})/H(\mathbb{Q})} \omega \in \mathbb{Q}$$

we conclude

$$\int_{H(\mathbb{R})/T} \omega_\infty \in \zeta_k(2) \cdots \zeta_k(j+1)\mathbb{Q}$$

and

$$\int_{H(\mathbb{R})/T} \omega \in i^{r_2(j^2+2j)} \zeta_k(2) \cdots \zeta_k(j+1)\mathbb{Q}.$$

This finishes the proof of Theorem 4.2.1. □

4.3. It remains to relate this computation to the continuous cohomology, and finally to the K-theory. We combine (3.4.4) and (4.1.1)

$$\begin{array}{ccc} H^*_{\mathrm{top}}(G_{n,u},\mathbb{C}) & \xleftarrow{\approx} & H^*_{\mathrm{top}}(K_n,\mathbb{C})\otimes H^*(X_{n,u},\mathbb{C}) \\ \wr\Big\uparrow \alpha^* & & \Vert\wr \Big\uparrow \mathrm{id}\otimes\bar\alpha^* \\ H^*(\mathfrak{g}_n,\mathbb{C}) & \xleftarrow{\approx} & H^*_{\mathrm{top}}(K_n,\mathbb{C})\otimes H^*(\mathfrak{g}_n,\mathfrak{K}_n;\mathbb{C}) \\ \wr\Big\downarrow \beta^* & & \Vert\wr \Big\downarrow \mathrm{id}\otimes\bar\beta^* \\ H^*(G_n(\mathbb{R})/\Gamma_n,\mathbb{C}) & \xleftarrow{\approx} & H^*_{\mathrm{top}}(K_n,\mathbb{C})\otimes H^*(X_m/\Gamma_n,\mathbb{C}) \end{array} \tag{4.3.1}$$

Let $\mu = \beta^*\alpha^{*-1}$, $\bar\mu = \bar\beta^*\bar\alpha^{*-1}$. By Proposition 3.5.1 and Theorem 4.2.1,

$$\mu L_m(G_{n,u},\mathbb{Q}) = \pi^{-d(m+1)} i^{d(m+1)+r_2}|D|^{1/2}\zeta_k(m-1)L_m(Y_n,\mathbb{Q}). \tag{4.3.2}$$

We have also isomorphisms

$$\begin{array}{ccc} H^*(\mathfrak{g}_n,\mathfrak{K}_n;\mathbb{C}) & \cong & H^*(X_n/\Gamma_n,\mathbb{C}) \\ \Vert\wr & & \Vert\wr \\ H^*_{\mathrm{cont}}(G_n(\mathbb{R}),\mathbb{C}) & \longrightarrow & H^*(\Gamma_n,\mathbb{C}) \end{array}$$

Viewing $\bar\mu$ as a map

$$H^*(X_{n,u},\mathbb{C}) \to H^*(\Gamma_n,\mathbb{C}),$$

and writing $I^{2m+1}(\Gamma_n,\mathbb{Q})$ for the quotient space of indecomposable elements,

$$L_m(\Gamma_n,\mathbb{Q}) = \bigwedge^{d_m} I^{2m+1}(\Gamma_n,\mathbb{Q})$$

(d_m as in Proposition 3.4.2), we find from (4.3.1)

$$\bar\mu L_m(X_{n,u},\mathbb{Q}) = (\pi i)^{-d(m+1)}|D|^{1/2} i^{r_2}\zeta_k(m+1)L_m(\Gamma_n,\mathbb{Q}). \tag{4.3.3}$$

REMARK 4.3.1. For k totally real and m odd, $d_m = 0$ so

$$I^{2m+1}(K_n,\mathbb{Q}) \cong I^{2m+1}(G_{n,u},\mathbb{Q}).$$

In this case μ is defined over $\mathbb{Q}$, so $\pi^{-d(m+1)}|D|^{1/2}\zeta_k(m+1) \in \mathbb{Q}$.

The map $\bar\mu\colon H^*(X_{n,u},\mathbb{C}) \to H^*(\Gamma_n,\mathbb{C})$ is not compatible with the real structures. Indeed, if $\mathfrak{g}_n(\mathbb{R}) = \mathfrak{K}_n \oplus \mathfrak{p}_n$, the tangent space to $X_{n,u}$ is $i\mathfrak{p}_n$, whereas that of X_n is $\mathfrak{p}_n$. It follows that the map $j_\Gamma = i^m\bar\mu$ in dimension m *is* compatible with the real structure. Rewriting Borel's result with j_Γ instead of $\bar\mu$,

$$j_\Gamma L_m(X_{n,u},\mathbb{Q}) = \pi^{-d(m+1)}|D|^{1/2}\zeta_k(m+1)L_m(\Gamma_n,\mathbb{Q}). \tag{4.3.4}$$

Let $\Gamma_n = \mathrm{SL}_n(\mathcal{O}_k)$, $\Gamma = \bigcup \mathrm{SL}_n(\mathcal{O}_k)$. Recall that $K_p(\mathcal{O}_k)$ is the pth homotopy group of an H-space $B^+_{\mathrm{GL}(\mathcal{O}_k)}$ with the same homology as $B_{\mathrm{GL}(\mathcal{O}_k)}$. There is, therefore, a Hurewitz map

$$K_p(\mathcal{O}_k) \to H_p(\Gamma,\mathbb{Z})$$

and $K_p(\mathcal{O}_k) \otimes \mathbb{Q}$ is identified with the dual of $I^p(\Gamma, \mathbb{Q})$. We can restate Borel's result as follows:

THEOREM 4.3.2. *$K_{2m}(\mathcal{O}_k)$ is torsion, and $K_{2m+1}(\mathcal{O}_k)$ has rank d_m. Moreover, the dual of the map j_Γ defined above (regulator map) embeds $K_{2m+1}(\mathcal{O}_k)/(\text{torsion})$ as a lattice of maximal rank in the primitive homology $I_{2m+1}(X_{n,u}, \mathbb{R})$. The volume of this lattice, computed with respect to a basis of $I_{2m+1}(X_{n,u}, \mathbb{Q})$ is a rational multiple of*

$$\pi^{-d(m+1)}|D|^{1/2}\zeta_k(m+1).$$

REMARK 4.3.3. The functional equation for ζ_k shows

$$\lim_{s \to -m} \zeta_k(s)(s+m)^{-d_m} \in \pi^{-d(m+1)+d_m}|D|^{1/2}\zeta_k(m+1)\mathbb{Q}$$

so the volume of the above lattice is a rational multiple of

$$\pi^{-d_m} \lim_{s \to -m} \zeta_k(s)(s+m)^{d_m}.$$

LECTURE 5

The Regulator Map. I

5.1. In the next three lectures, we will write down explicit formulas for the regulator map

$$D\colon K_3(\mathbb{C}) \to \mathbb{R}.$$

These formulas will subsequently be generalized to the case of an elliptic curve over $\mathbb{C}$.

We show in Lecture 7 that on the level of cohomology D gives a measurable 3-cocycle on $\mathrm{SL}_2(\mathbb{C})$, whose class in $H^3_{\mathrm{cont}}(\mathrm{SL}_2(\mathbb{C}), \mathbb{R})$ is

$$\frac{2}{3} \times (\text{class of volume form on } \mathrm{SU}_2 \setminus \mathrm{SL}_2(\mathbb{C})).$$

Also in Lecture 7 for $k = \mathbb{Q}(\zeta)$, $\zeta^\ell = 1$ prime cyclotomic, we will associate to each ζ^i an element $[\zeta^i] \in K_3(k)/\text{torsion}$. We will show $[\zeta^i] + [\zeta^{-i}] = 0$ and we will calculate explicitly the regulator associated to the basis $[\zeta]$, $[\zeta^2]$, ..., $[\zeta^{(ell-1)/2}]$ of $K_3(k) \otimes \mathbb{Q}$, comparing it to $\zeta_k(2)$. (Note that unlike the classical situation for units, one has $K_3(\mathcal{O}_k) \twoheadrightarrow K_3(k)$ with torsion kernel so "integrality" plays no role.) The calculation will also be later generalized to the context of elliptic curves.

5.2. We begin by recalling some algebraic K-theory which will be useful. Let C be a smooth projective curve defined over an algebraically closed field k. Let $T \subset C$ be a finite set of points (possibly empty). $K_*(C)$ (resp. $K_*(T)$) will denote the homotopy groups of the classifying space of the Quillen category of locally-free sheaves on C (resp. on $T = \coprod_T \mathrm{Spec}(k)$).

$$K_*(C) = \pi_{*_{+1}} BQ_C$$
$$K_*(T) = \pi_{*_{+1}} BQ_T.$$

$K_*(C,T)$ will denote the homotopy of the fibre $F_{C,T}$ of the map $BQ_C \to BQ_T$,

$$K_*(C,T) = \pi_{*_{+1}} F_{C,T}.$$

Similarly we write C_T for the spectrum of the semi-local ring of functions on C regular at all points of T, and define $K_*(C_T)$ and $K_*(C_T, T)$.

We define

$$SK_*(C,T) \overset{\text{dfn}}{=} \operatorname{Ker}\bigl(K_*(C,T) \xrightarrow{\text{restrict}} K_*(C_T,T)\bigr).$$

PROPOSITION 5.2.1. *There is a relative localization sequence*

$$\cdots \to K_n(C,T) \to K_n(C_T,T) \to \bigoplus_{p\in C-T} K_{n-1}\bigl(k(p)\bigr) \to K_{n-1}(C,T) \to \cdots$$

Of particular interest to us is the following special case:

COROLLARY 5.2.2. *There is an exact sequence*

$$0 \to K_2(C,T)/SK_2(C,T) \to K_2(C_T,T) \to \bigoplus_{C-T} k^* \to SK_1(C,T) \to 0$$

The proposition follows from a more general result.

PROPOSITION 5.2.3. *Let X be a regular noetherian scheme, and let Y, $Z \subset X$ be closed subschemes. Assume $Y \cap Z = \varnothing$. Then there is a long exact localization sequence*

$$\cdots \to K_n(X,Z) \to K_n(X-Y,Z) \to K'_{n-1}(Y) \to K_{n-1}(X,Z) \to \cdots$$

where $K'_n(Y)$ denotes the K-theory of the category of coherent $\mathcal{O}_Y$-modules.

PROOF. By a theorem of Quillen, the fibre $H_{X,Y}$ of $BQ_X \to BQ_{X-Y}$ has homotopy groups $K'_*(Y)$. We consider the diagram of fibrations

$$\begin{array}{ccccc}
F_{X-Y,Z} & \longrightarrow & BQ_{X-Y} & \longrightarrow & BQ_Z \\
\uparrow & & \uparrow & & \| \\
F_{X,Z} & \longrightarrow & BQ_X & \longrightarrow & BQ_Z \\
 & & \uparrow & & \uparrow \\
 & & H_{X,Y} & \longrightarrow & *.
\end{array}$$

and deduce a fibration $H_{X,Y} \to F_{X,Z} \to F_{X-Y,Z}$. The corresponding homotopy sequence is the desired one. □

EXAMPLE 5.2.4. Suppose $T = \varnothing$. By a theorem of Quillen, there is a resolution of sheaves

$$0 \to \underline{\underline{K}}_{2_C} \to i_{\eta^*}\left(K_2\bigl(k(C)\bigr)\right) \to \coprod_{p\in C} i_{p^*} k(p)^* \to 0$$

where the left hand sheaf is the Zariski sheaf associated to K_2 on C, the middle is the constant sheaf with value $K_2\bigl(k(C)\bigr)$, and the right hand is the skyscraper sheaf with stalk $k(p)^*$ at p. Taking cohomology, one finds

$$0 \to \Gamma(C, \underline{\underline{K}}_2) \to K_2\big(k(C)\big) \xrightarrow{\text{tame}} \coprod_p k(p)^* \to H^1(C, \underline{\underline{K}}_2) \to 0.$$

Comparison with Corollary 5.2.2 shows

$$SK_1(C) \cong H^1(C, \underline{\underline{K}}_2)$$
$$K_2(C)/SK_2(C) \cong \Gamma(C, \underline{\underline{K}}_2).$$

In fact we have similar isomorphisms for all the K_i.

EXAMPLE 5.2.5. Let $C = \mathbb{P}^1_k$, $T = \{0, \infty\}$. By yet another fundamental result of Quillen

$$\begin{aligned} K_n(\mathbb{P}^1_k) &\cong K_n(k) \otimes_{\mathbb{Z}} K_0(\mathbb{P}^1) \\ &\cong K_n(k) \cdot [\mathcal{O}_{\mathbb{P}^1}] \oplus K_n(k) \cdot [k(\text{point})]. \end{aligned}$$

The restriction map $K_n(\mathbb{P}^1_k) \to K_n(\{0, \infty\}) \cong K_n(k) \oplus K_n(k)$ is zero on $K_n(k) \cdot [k(\text{point})]$ and maps the factor $K_n(k) \cdot [\mathcal{O}_{\mathbb{P}^1}]$ diagonally. From this one deduces an isomorphism

$$K_n(\mathbb{P}^1_k, \{0, \infty\}) \cong K_{n+1}(k) \oplus K_n(k) \tag{5.2.1}$$

as follows: the sequence

$$K_{n+1}(\mathbb{P}^1) \to K_{n+1}(\{0, \infty\}) \to K_n(\mathbb{P}^1, \{0, \infty\}) \to K_n(\mathbb{P}^1) \to K_n(\{0, \infty\})$$

yields

$$0 \to K_{n+1}(k) \to K_n(\mathbb{P}^1, \{0, \infty\}) \to K_n(k) \to 0.$$

But $K_*(\mathbb{P}^1, \{0, \infty\})$ will have a K_*-module structure, so we can split the above by mapping $\alpha \in K_n(k)$ to $\alpha \cdot x$, where $x \in K_0(\mathbb{P}^1, \{0, \infty\})$ is the class of the residue field $k(1)$ at the point $1 \in \mathbb{P}^1 \setminus \{0, \infty\}$.

We now analyze the localization sequence multlinegap0pt

$$\begin{aligned} \cdots \to \coprod_{p \in \mathbb{P}^1 \setminus \{0,\infty\}} K_n(k) &\to K_n(\mathbb{P}^1, \{0, \infty\}) \\ &\to K_n(\mathbb{P}^1_{\{0,\infty\}}, \{0, \infty\}) \to \coprod K_{n-1}(k) \to \cdots . \end{aligned} \tag{5.2.2}$$

For $n = 0$, the map

$$K_0\big(k(p)\big) \cong \mathbb{Z} \to K_0(\mathbb{P}^1, \{0, \infty\})$$

sends 1 to the class of $k(p)$. Letting $\coprod^{\circ} \mathbb{Z}$ denote the kernel of the "degree map"

$$\coprod_{p \in \mathbb{P}^1 \setminus \{0,\infty\}} \mathbb{Z} \to \mathbb{Z},$$

we obtain from (5.2.1) a map $\rho\colon \coprod^{\circ} \mathbb{Z} \to k^*$ which can be described as follows: given $(\dots a_p \dots) \in \coprod^{\circ} \mathbb{Z}$ there exists a function f on $\mathbb{P}^1$ such that $(f) = \sum a_p(p)$. Then

$$\rho(\dots a_p \dots) = f(0)f(\infty)^{-1} \in k^*.$$

Clearly ρ is surjective. For arbitrary n we use the compatibility of the localization sequence with the $K_*(k)$-module structure to deduce

$$\begin{aligned}\operatorname{Coker}\Big(\coprod_{\mathbb{P}^1\setminus\{0,\infty\}} K_n(k) \to K_n(\mathbb{P}^1, \{0,\infty\})\Big)& \\ \cong \operatorname{Coker}\big(K_n(k)\otimes k^* &\xrightarrow{\text{multiply}} K_{n+1}(k)\big) \\ \overset{\text{dfn}}{=} K_{n+1}(k)^{\text{indecomposable}}&.\end{aligned}$$

We now go back to the localization sequence and consider the case $n = 2$. Since $K_2(k)$ is known to be generated by symbols $K_2(k)^{\text{ind}} = (0)$, the analogue of Example 5.2.4 becomes

$$\begin{aligned}(5.2.3)\quad 0 \to K_3(k)^{\text{ind}} &\to K_2(\mathbb{P}^1_{\{0,\infty\}}, \{0,\infty\}) \\ &\xrightarrow{\text{tame}} \coprod_{p\neq 0,\infty} k^*_{(p)} \to K_2(k) \oplus k^* \to 0.\end{aligned}$$

Finally, let $\coprod^{\circ} k(p)^*$ denote the kernel of the obvious norm mapping $\coprod k(p)^* \to k^*$ both for Example 5.2.4 and (5.2.3). In both cases the image of the tame mapping lands in $\coprod^{\circ} k(p)^*$, so we get

$$\begin{aligned}(5.2.4)\qquad 0 \to \Gamma(C, \underline{\underline{K}}_2) &\to K_2\big(k(C)\big) \\ &\xrightarrow{\text{tame}} \coprod_{p\in C}^{\circ} k(p)^* \to H^1(C, \underline{\underline{K}}_2)^{\circ} \to 0\end{aligned}$$

$$\begin{aligned}(5.2.5)\qquad 0 \to K_3(k)^{\text{ind}} &\to K_2(\mathbb{P}^1_{\{0,\infty\}}, \{0,\infty\}) \\ &\xrightarrow{\text{tame}} \coprod_{p\neq 0,\infty}^{\circ} k(p)^* \to K_2(k) \to 0.\end{aligned}$$

We stress again the analogy between $K_{2n+1}(k)^{\text{ind}}$ and

$$K_{2n}(k)^{\text{decomposable}} \overset{\text{dfn}}{=} \operatorname{Image}\big(K_{2n-1}(k)\otimes k^* \to K_{2n}(k)\big)$$

on the one hand, and $\Gamma(C, \underline{\underline{K}}_{2n})$ and $SK_{2n-1}(C) = H^1(C, \underline{\underline{K}}_{2n})$ on the other hand.

5.3. To simplify notation, let R denote the semi-local ring at 0 and ∞ on $\mathbb{P}^1$, and let $I \subset R$ be the ideal of $\{0, \infty\}$. When we want to specify a particular ground field k, we write R_k, I_k. For now, we take $k = \mathbb{C}$. We define a map

$$(1+I)^* \otimes_{\mathbb{Z}} k(\mathbb{P}^1)^* \to \coprod_{\substack{x\in\mathbb{P}^1 \\ x\neq 0,\infty}} \mathbb{Z} = \coprod_{x\in\mathbb{C}^*} \mathbb{Z} \tag{5.3.1}$$
$$f \otimes g \longmapsto (f^-) * (g)$$

as follows: Let t be the standard parameter on $\mathbb{P}^1$ ($t(a) = a$, $a = 0$, 1, ∞). Since $f(0) = f(\infty) = 1$, we can write

$$f = \prod (t-\alpha_i)^{d_i}, \quad \sum d_i = 0, \quad \prod \alpha_i^{d_i} = 1.$$

Let $f^- = \prod (t - \alpha_i^{-1})^{d_i}$. Write

$$g = c\prod (t-\beta_j)^{e_j}, \qquad c \in \mathbb{C}, \beta_j \in \mathbb{C}.$$
$$e = -\sum e_j = \operatorname{ord}_\infty g.$$

Finally let

$$(f) = \sum d_i(\alpha_i), \quad (f^-) = \sum d_i(\alpha_i^{-1})$$

denote the divisors of these functions. We take

$$(f^-) * (g) = \sum_{\substack{i,j \\ \beta_j \neq 0}} d_i e_j (\alpha_i^{-1}\beta_j) + e(f) + e(f^-). \tag{5.3.2}$$

We remark now for future use that there is a simpler analogue of this construction for elliptic curves E. Let $k(E)$ denote the function field of E (over some algebraically closed field k) and define

$$k(E)^* \otimes k(E)^* \to \coprod_{p\in E} \mathbb{Z} \tag{5.3.3}$$
$$f \otimes g \longmapsto (f)^- * (g)$$

where if $(f) = \sum d_i(a_i)$, $(g) = \sum e_j(b_j)$ we have

$$(f)^- = \sum d_i(-a_i), \quad (f)^- * (g) = \sum d_i e_j (b_j - a_i). \tag{5.3.4}$$

From work of Keune (see below) and from the well-known presentation of K_2 of a field due to Matsumoto, we are led to consider the symbols

$$f \otimes (1-f) \in (1+I)^* \otimes \mathbb{C}(\mathbb{P}^1)^* \qquad \text{for } f \in 1+I,\ f \not\equiv 1$$
$$f \otimes (1-f) \in \mathbb{C}(E)^* \otimes \mathbb{C}(E)^* \qquad \text{for } f \in \mathbb{C}(E)^*,\ f \not\equiv 1.$$

We suppose now given a function $D\colon \mathbb{C}^* \to \mathbb{R}$ (resp. $D_E\colon E \to \mathbb{R}$).

DEFINITION 5.3.1. The function D (resp. D_E) induces a homomorphism $\coprod_{\mathbb{C}^*}\mathbb{Z}\to\mathbb{R}$ (resp. $\coprod_E\mathbb{Z}\to\mathbb{R}$). D (resp. D_E) is called a *relative Steinberg function* (resp. *Steinberg function*) for $\{0,\infty\}\subset\mathbb{P}^1_{\mathbb{C}}$ (resp. for E) if

$$D\big((f)^- * (1-f)\big)=0,\quad \text{all } f\in 1+I,\ f\not\equiv 1$$

$$(\text{resp. } D\big((f)^- * (1-f)\big)=0,\quad \text{all } f\in\mathbb{C}(E)^*,\ f\not\equiv 1.$$

PROPOSITION 5.3.2. *Let*

$$\mathcal{R}\subset(1+I)^*\otimes\mathbb{C}(\mathbb{P}^1)^*\quad (\textit{resp. } \mathcal{R}_E\subset\mathbb{C}(E)^*\otimes\mathbb{C}(E)^*)$$

denote the subgroup generated by the symbols $f\otimes(1-f)$. Then there is a natural homomorphism

$$K_2(R,I)\to\frac{(1+I)^*\otimes\mathbb{C}(\mathbb{P}^1)^*}{\mathcal{R}}$$

(*resp. an isomorphism*

$$K_2\big(\mathbb{C}(E)\big)\to\frac{\mathbb{C}(E)^*\otimes\mathbb{C}(E)^*}{\mathcal{R}_E}).$$

COROLLARY 5.3.3. *A relative Steinberg function* (*resp. a Steinberg function on E*) *induces a homomorphism*

$$K_2(R,I)\to\mathbb{R}\quad (\textit{resp. } K_2\big(\mathbb{C}(E)\big)\to\mathbb{R}).$$

PROOF OF PROPOSITION 5.3.2. The assertion for E is exactly the well-known Matsumoto presentation for K_2 of a field. In the relative case we use a presentation due to Keune [**Keu78**] and valid for $K_2(R,I)$ for any semi-local ring R and any ideal I in the Jacobson radical. Generators are elements $\langle a,b\rangle$ with $a\in R,\ b\in I$, or $a\in I,\ b\in R$. (Following a suggestion of Stienstra, we change Keune's notation slightly so our $\langle a,b\rangle$ = Keune's $\langle -a,b\rangle$.) Relations are

$$\langle a,b\rangle+\langle b,a\rangle=0 \tag{5.3.5}$$

$$\langle a,b\rangle+\langle a,c\rangle=\langle a,b+c-abc\rangle \tag{5.3.6}$$

$$\langle a,bc\rangle=\langle ab,c\rangle+\langle ac,b\rangle \tag{5.3.7}$$

where all symbols are assumed to make sense, i.e., to have at least one element in I.

We map

$$K_2(R,I)\to\frac{(1+I)^*\otimes\mathbb{C}(\mathbb{P}^1)^*}{\mathcal{R}}$$

$$\langle a,b\rangle\mapsto(1-ab)\otimes b\equiv(1-ab)\otimes a^{-1}$$

so

$$\begin{aligned}
\langle a,b\rangle + \langle b,a\rangle &\to (1-ab)\otimes ab \equiv 1\otimes 1\\
\langle a,b\rangle + \langle a,c\rangle &\to \big((1-ab)\otimes b\big) + \big((1-ac)\otimes c\big)\\
&\equiv \big((1-ab)\otimes a^{-1}\big) + \big((1-ac)\otimes a^{-1}\big)\\
&= (1-ab-ac+a^2bc)\otimes a^{-1}\\
&\equiv (1-ab-ac+a^2bc)\otimes(-b+c-abc)\\
&= \text{Image of } \langle a, b+c-abc\rangle\\
\langle a,bc\rangle \mapsto (1-abc)\otimes bc &= \big((1-abc)\otimes b\big) + \big((1-abc)\otimes c\big)\\
&= \text{Image of } \langle ac,b\rangle + \langle ab,c\rangle.
\end{aligned}$$

□

REMARK 5.3.4. It seems likely that

$$K_2(R,I) \cong \frac{(1+I)^* \otimes \mathbb{C}(\mathbb{P}^1)^*}{\mathcal{R}}$$

but I have not pursued this point.

LECTURE 6

The Dilogarithm Function

6.1. The purpose of this lecture is to introduce the dilogarithm function and to derive from it a relative Steinberg function $D(x)$, following [**Blo78**]. For an Abelian group A, $\tilde{A}$ will denote $A/\text{torsion}$. We have an exact sequence

$$0 \to \tilde{\mathbb{C}}^* \xrightarrow{1\otimes \mathrm{Id}} \mathbb{C} \otimes \mathbb{C}^* \xrightarrow{\exp(2\pi i \cdot)\otimes \mathrm{Id}} \mathbb{C}^* \otimes_{\mathbb{Z}} \mathbb{C}^* \to 0 \tag{6.1.1}$$

obtained by tensoring the standard exponential sequence

$$0 \to \mathbb{Z} \xrightarrow{1} \mathbb{C} \xrightarrow{\exp(2\pi i \cdot)} \mathbb{C}^* to 0$$

with $\mathbb{C}^*$. We have also for any field F an exact sequence

$$0 \to \mathcal{B}(F) \to \mathcal{A}(F) \xrightarrow{\lambda} F^* \otimes_{\mathbb{Z}} F^* \xrightarrow{S} K_2(F) \to 0 \tag{6.1.2}$$

where $\mathcal{A}(F)$ = free Abelian group on generators $[f]$, $f \in F$, $f \neq 0, 1$, and $\lambda[f] = (1-f) \otimes f$. Of course $\mathcal{B}(f) = \operatorname{Ker} \lambda$.

LEMMA 6.1.1. *For $a \in \mathbb{C} \setminus \{0,1\}$, the expression*

$$\epsilon[a] \stackrel{\text{dfn}}{=} \left[\frac{1}{2\pi i} \log(1-a) \otimes a\right] + \left[1 \otimes \exp\left(\frac{-1}{2\pi i} \int_0^a \log(1-t) \frac{dt}{t}\right)\right] \in \mathbb{C} \otimes_{\mathbb{Z}} \mathbb{C}^*$$

is well-defined independent of the choices of branches for the multivalued functions, so we get a commutative triangle

$$\begin{array}{ccc} & \mathcal{A}(\mathbb{C}) & \\ {}^{\epsilon}\swarrow & & \searrow^{\lambda} \\ \mathbb{C} \otimes_{\mathbb{Z}} \mathbb{C}^* & \xrightarrow[\exp(2\pi i \cdot)\otimes \mathrm{Id}]{} & \mathbb{C}^* \otimes \mathbb{C}^* \end{array} \tag{6.1.3}$$

PROOF. The assertion of the lemma should be interpreted as follows: a value for the functions $\log(1-a)$ and $\int_0^a \log(1-t)dt/t$, and hence for $\epsilon[a]$, is determined by the choice of a path γ from 0 to a. The expression $\epsilon[a]$ is independent of γ. To see this, the key configuration to be considered looks like

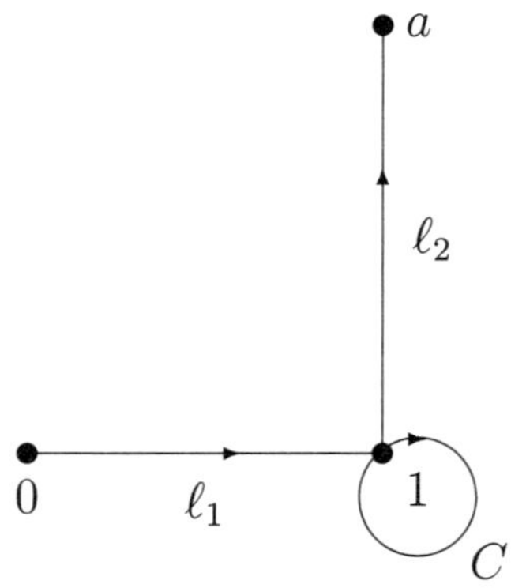

where C denotes a small circle about 1, and the two paths are $\gamma = \ell_1 + \ell_2$ and $\gamma' = \ell_1 + C + \ell_2$. We have

$$\left[\frac{1}{2\pi i}\log(1-a)\otimes a\right]_{\substack{\text{eval.}\\ \text{via } \gamma'}} - \left[\frac{1}{2\pi i}\log(1-a)\otimes a\right]_{\substack{\text{eval.}\\ \text{via } \gamma'}} = 1 \otimes a$$

$$\left[1\otimes\exp\left(\frac{-1}{2\pi i}\int_0^a \log(1-t)\frac{dt}{t}\right)\right]_{\substack{\text{eval.}\\ \text{via } \gamma'}} - \left[1\otimes\exp\left(\frac{-1}{2\pi i}\int_0^a \log(1-t)\frac{dt}{t}\right)\right]_{\substack{\text{eval.}\\ \text{via } \gamma'}}$$

$$\xrightarrow{\text{as } C\to 0} 1\otimes\exp\left(\frac{-1}{2\pi i}\int_0^a 2\pi i\frac{dt}{t}\right) = 1\otimes a^{-1}.$$

One must still consider possible problems at 0 due to the fact that $\log(1-t)\,dt/t$ has a residue at 0 for a non-principal branch of the logarithm. However if C' is a small circle about 0 and $\log(1-t) = 2n\pi i +$ principal branch $\log(1-t)$ for t inside C' we get

$$\exp\left(\frac{-1}{2\pi i}\int_{C'}\log(1-t)\frac{dt}{t}\right) = \exp\left(-n\int_{C'}\frac{dt}{t}\right) = 1$$

so these problems disappear as well. □

COROLLARY 6.1.2. *The expression*

$$D(x) = \arg(1-x)\log|x| - \operatorname{Im}\left(\int_0^x \log(1-t)\frac{dt}{t}\right)$$

gives a well-defined function $\mathbb{C}\to\mathbb{R}$.

PROOF. Consider the diagram (defining Φ)

$$
\begin{array}{ccccccc}
0 & \longrightarrow & \mathcal{B}(\mathbb{C}) & \longrightarrow & \mathcal{A}(\mathbb{C}) & \xrightarrow{\lambda} & \mathbb{C}^* \otimes \mathbb{C}^* \\
 & & \downarrow \Phi & & \downarrow \epsilon & & \| \\
0 & \longrightarrow & \widetilde{\mathbb{C}}^* & \longrightarrow & \mathbb{C} \otimes \mathbb{C}^* & \longrightarrow & \mathbb{C}^* \otimes \mathbb{C}^* \\
 & & \Big| & & \downarrow (\text{real part})\otimes \text{Id} & & \\
 & & \Big| \log|\ | & & \mathbb{R} \otimes \mathbb{C}^* & & \\
 & & \downarrow & & \downarrow \text{Id}\otimes\log|| & & \\
 & & \mathbb{R} & \xleftarrow[\text{multiply}]{} & \mathbb{R}\otimes\mathbb{R} & &
\end{array}
\tag{6.1.4}
$$

The composition

$$(\text{multiply}) \circ (\text{Id} \otimes \log|\ |) \circ \big((\text{real part}) \otimes \text{Id}\big) \circ \epsilon \colon \mathcal{A}(\mathbb{C}) \to \mathbb{R}$$

is easily seen to be

$$[a] \mapsto D(a),$$

so D is well-defined on $\mathbb{C} \setminus \{0,1\}$. Taking $D(0) = D(1) = 0$ gives a continuous extension to all of $\mathbb{C}$. $\square$

6.2. This construction has an important rigidity property. To formulate it, let A denote the ring of holomorphic functions, say, on some fixed disk about 0 in $\mathbb{C}$. Let $A^0 \subset A$ denote the subset of functions $a(z)$ such that $a(z)$ and $1 - a(z) \in A^*$. Let $\mathcal{A}(A) =$ free Abelian group on generators $[a(z)]$, $a(z) \in A^0$, and define $\mathcal{B}(A)$ by the exact sequence

$$0 \to \mathcal{B}(A) \to \mathcal{A}(A) \xrightarrow{\lambda} A^* \otimes_{\mathbb{Z}} A^*$$
$$\lambda[a(z)] = \big(1 - a(z)\big) \otimes a(z).$$

Let Ω^1_A denote the A-module of holomorphic 1-forms on the disk, and consider the commutative diagram

$$
\begin{array}{ccccc}
\mathcal{B}(A) & \hookrightarrow & & \longrightarrow & \mathcal{A}(A) \\
\downarrow \Phi_A & & & & \downarrow \epsilon_A \\
\tilde{A}^* & \hookrightarrow & \xrightarrow{1\otimes\text{Id}} & \longrightarrow & A \otimes_{\mathbb{Z}} A^* \\
 & \searrow^{\text{dlog}} & & \swarrow_{\text{Id}\otimes\text{dlog}} & \\
 & & \Omega^1_A & &
\end{array}
$$

where Φ_A and ϵ_A are defined precisely as Φ, ϵ above, and

$$(\text{Id} \otimes \text{dlog})(a \otimes b) \stackrel{\text{dfn}}{=} a \frac{db}{b}.$$

But

$$(\mathrm{Id}\otimes \mathrm{dlog})\epsilon_A([a(z)])$$
$$= (\mathrm{Id}\otimes \mathrm{dlog})\left\{\left(\frac{1}{2\pi i}\log(1-a)\otimes a\right)+\left(1\otimes \exp\left(\frac{-1}{2\pi i}\int_0^{a(z)}\log(1-t)\frac{dt}{t}\right)\right)\right\}$$
$$= \frac{1}{2\pi i}\log\bigl(1-a(z)\bigr)\frac{a'(z)}{a(z)} - \frac{1}{2\pi i}\log\bigl(1-a(z)\bigr)\frac{a'(z)}{a(z)} = 0.$$

This implies:

LEMMA 6.2.1. *With notation as above,* $\Phi_A\bigl(\mathcal{B}(A)\bigr) \subseteq \widetilde{\mathbb{C}}^* \subset \tilde{A}^*$.

COROLLARY 6.2.2. *Let* $D(X)$ *be as in Corollary* 6.1.2 *and suppose* $\sum n_i[a_i(z)] \in \mathcal{B}(A)$ *(i.e.,* $\prod\bigl[\bigl(1-a_i(z)\bigr)\otimes a_i(z)\bigr]^{n_i} = e$ *in* $A^*\otimes A^*$*). Then* $\sum n_i D\bigl(a_i(z)\bigr) \equiv \mathrm{const}$.

Let f_z denote an analytic family of meromorphic functions on $\mathbb{P}^1$ parametrized by z in some small disk about 0. We assume that neither f_z nor $1-f_z$ has multiple roots for any z, and that moreover $f_z(0)\equiv f_z(\infty)\equiv 0$. We may then write

$$f_z = c(z)\prod_{j\in J}\bigl(t-\beta_j(z)\bigr)^{e_j}$$
$$1-f_z = \prod_{i\in I}\bigl(t-\alpha_i(z)\bigr)^{d_i} \qquad \text{with } \sum d_i = 0,\ \prod \alpha_i(z)^{d_i} = 1$$

where t is the standard parameter on $\mathbb{P}^1$, and d_i, $e_j = \pm 1$. Note that either $\alpha_i(z)^{-1}\beta_j(z)\equiv 1$, or $\alpha_i(z)^{-1}\beta_j(z)\neq 1$ for any z. The expression

$$(1-f_z)^- * (f_z) = \sum_{\substack{i,j\\ \alpha_i\neq\beta_j\neq 0}} d_i e_j[\alpha_i(z)^{-1}\beta_j(z)] + e\sum_i d_i([\alpha_i(z)]+[\alpha_i(z)^{-1}]),$$

with $e = -\sum e_j$ therefore makes sense in $\mathcal{A}(A)$ as long as no $\alpha_i(z) = 1$.

LEMMA 6.2.3. *Let* f_z *be as above. Then* $(1-f_z)^- * (f_z) \in \mathcal{B}(A)$.

PROOF. Let I, J be the index sets for the roots of $1-f_z$, f_z respectively. Let $K\subset I$, J be the common set of poles. For $k\in I$ we have $\alpha_k(z) = \beta_k(z)$, $d_k = e_k = -1$. Write $I = K \amalg J'$, $J = K\amalg J'$. Then

$$\begin{aligned}
\lambda&\big((1-f_z)^- * (f_z)\big)\\
&= \prod_{\substack{i\in I'\\ j\in J}}\big((1-\alpha_i^{-1}\beta_j)^{d_ie_j}\otimes\alpha_i^{-1}\big) \prod_{\substack{i\in I, j\in J'\\ \beta_j\neq 0}}\big((1-\alpha_i^{-1}\beta_j)^{d_ie_j}\otimes\beta_j\big)\\
&\quad\times \prod_{\substack{k\in K, j\in J\\ j\neq k}}\big((1-\alpha_k^{-1}\beta_j)^{d_ke_j}\otimes\alpha_k^{-1}\big) \prod_{\substack{k\in K, i\in I\\ i\neq k}}\big((1-\alpha_i^{-1}\alpha_k)^{d_id_k}\otimes\alpha_k\big)\\
&\qquad\times \prod_{i\in I}\big((-\alpha_i)^{d_i}\otimes\alpha_i\big)^e
\end{aligned}$$

(We use here that $\big((1-\alpha)\otimes\alpha\big)\big((1-\alpha^{-1})\otimes\alpha^{-1}\big) = (-\alpha)\otimes\alpha$.) We evaluate these products

$$\begin{aligned}
\text{(6.2.1)}\quad \prod_{\substack{i\in I'\\ j\in J}}&\big((1-\alpha_i^{-1}\beta_j)^{d_ie_j}\otimes\alpha_i^{-1}\big)\\
&= \prod_{i\in I'}(\alpha_i^{d_ie}\otimes\alpha_i^{-1})\cdot\prod_{\substack{i\in I'\\ j\in J}}\big((\alpha_i-\beta_j)^{d_ie_j}\otimes\alpha_i^{-1}\big)\\
&= \prod_{i\in I'}(\alpha_i^{d_ie}\otimes\alpha_i^{-1})\prod_{i\in I'}(c\otimes\alpha_I^{d_i})
\end{aligned}$$

because for $i\in I'$, $f(\alpha_i)=1$ so

$$\prod_{j\in J}(\alpha_i-\beta_j)^{d_ie_j} = c^{-d_i}f(\alpha_i)^{d_i} = c^{-d_i}.$$

Next

$$\text{(6.2.2)}\quad \prod_{\substack{i\in I, j\in J'\\ \beta_j\neq 0}}\big((1-\alpha_i^{-1}\beta_j)^{d_ie_j}\otimes\beta_j\big) = \prod_{\substack{j\in J'\\ \beta_j\neq 0}}\left(\prod_{i\in I}(\beta_j-\alpha_i)^{d_ie_j}\otimes\beta_j\right) = 1.$$

Also

$$\begin{aligned}
\text{(6.2.3)}\quad &\prod_{\substack{k\in K, j\in J\\ j\neq k}}\big((1-\alpha_k^{-1}\beta_j)^{d_ke_j}\otimes\alpha_k^{-1}\big) \prod_{\substack{k\in K, i\in I\\ i\neq k}}\big((1-\alpha_i^{-1}\alpha_k)^{d_id_k}\otimes\alpha_k\big)\\
&= \prod_{k\in K}\left\{\frac{\prod_{\substack{i\in I\\ i\neq k}}(-\alpha_i)^{-d_id_k}\prod_{\substack{i\in I\\ i\neq k}}(\alpha_k-\alpha_i)^{d_id_k}}{\alpha_k^{d_k(e+d_k)}\prod_{\substack{j\in J\\ j\neq k}}(\alpha_k-\beta_j)^{d_ke_j}}\otimes\alpha_k\right\}\\
&= \prod_{k\in K}\left\{\left(\frac{(-\alpha_k)^{d_k^2}}{\alpha_k^{d_ke+d_k^2}}\cdot c^{d_k}\lim_{x\to\alpha_k}\left(\frac{1-f_z(x)}{f_z(x)}\right)^{d_k}\right)\otimes\alpha_k\right\}\\
&= \prod_{k\in K}(\alpha_k^{d_ke}\otimes\alpha_k^{-1})(c\otimes\alpha_k^{d_k}).
\end{aligned}$$

Combining (6.2.1)–(6.2.3) yields

$$\lambda\big((1-f_z)^- * (f_z)\big) = 1. \qquad \square$$

THEOREM 6.2.4. *The function* $D(x)$ *in Corollary* 6.1.2 *is a relative Steinberg function* (*cf. Definition* 5.3.1).

PROOF. Given $f, g \in I$, one can find an analytic family f_z with $f_0 = f$, $f_1 = g$. Moreover one can arrange that except for a finite number of values of z, say $z_1, \dots, z_n$, the functions f_z, $1 - f_z$ have no multiple zeros or poles, and no zero or pole of $1 - f_z$ is 1. It follows from Corollary 6.2.2 and Lemma 6.2.3 that

$$D\big((1-f_z)^- * (f_z)\big) \equiv \text{const.}, \quad z \neq z_1, \dots, z_n. \tag{6.2.4}$$

We claim, however, that the function of z on the left is defined and continuous at $z = z_i$. Indeed $D(0) = D(1) = 0$ so we may simply write (notation as above)

$$D\big((1-f_z) * (f_z)\big) = \sum_{i,j} d_i e_j D(\alpha_i^{-1}\beta_j) + e\sum_i d_i\big(D(\alpha_i) + D(\alpha_i^{-1})\big).$$

We will see below (Lemma 6.2.5) however that $D(x) = -D(x^{-1})$ so the second sum in the above expression drops out. Also we can extend D to a continuous function on all of $\mathbb{P}^1_{\mathbb{C}}$ by taking $D(\infty) = D(0) = 0$. Thus

$$D\big((1-f_z)^- * (f_z)\big) = \sum_{i,j} d_i e_j D(\alpha_I^{-1}\beta_j) \tag{6.2.5}$$

where the β_j run through *all* zeros and poles of f_z, including 0 and ∞. This expression is clearly continuous in z.

It remains to show that the constant in (6.2.4) is zero. Fix $f \in I$ and take $f_z = c(z)f$ where $c(z) \to 0$. The β_j are then fixed, but the zeros and poles of $1 - f_z$ coalesce. It follows from (6.2.5) that $D\big((1-f_z)^- * (f_z)\big) \to 0$. $\square$

We used in the above the following:

LEMMA 6.2.5. *Let* D *be as in Corollary* 6.1.2. *Then*

$$D(x) = -D(x^{-1}).$$

PROOF. We have

$$\log(1-x)\frac{dx}{x} + \log\left(1 - \frac{1}{x}\right)\frac{d(1/x)}{1/x} = \log(-x)\frac{dx}{x}.$$

Note

$$\operatorname{Im}\left(\int^x \log(-t)\frac{dt}{t}\right) = \operatorname{Im}\left(\frac{1}{2}\big(\log(-x)\big)^2\right) + C + C'\log|x|$$

where C' accounts for the ambiguity in the branch of $\log(-t)$. Thus

$$D(x)+D(x^{-1}) = -\frac{1}{2}\operatorname{Im}\left(\left(\log(-x)\right)^2\right)+\arg(-x)\log|x|+C+C''\log|x|$$

with C'' accounting for the ambiguity in the branches of $\log(-t)$ and $\arg(1-x)$. From this we see

$$D(x)+D(x^{-1}) = C + C''\log|x|.$$

However for $x<0$ it is clear from the definition that $D(x)=0$. Thus $C+C''\log|x| = 0$ for all $x<0$, whence $C=C''=0$. □

We note also, while we are at it

LEMMA 6.2.6. (i) $D(\bar{x}) = -D(x)$. *In particular* $D\equiv 0$ *on* $\mathbb{R}$.

(ii) $D(1-x) = -D(x)$.

PROOF. (i) is straightforward and won't be used anyway, so we leave it to the reader. For (ii), integrate by parts to get

$$D(x) = \arg(1-x)\log|x| - \operatorname{Im}\left[\log(1-x)\log(x) + \int_0^x \log t\frac{dt}{1-t}\right].$$

Take $x = 1-u$ in the integral and note $\int_0^1 \log(1-u)\,du/u$ is real:

$$\begin{aligned}D(x) &= \arg(1-x)\log|x| - \operatorname{Im}\left[\log(1-x)\log(x) - \int_0^{1-x}\log(1-u)\frac{du}{u}\right]\\ &= -\arg(x)\log|1-x| + \operatorname{Im}\int_0^{1-x}\log(1-u)\frac{du}{u}\\ &= -D(1-x).\end{aligned}$$

□

REMARK 6.2.7. From (6.2.5) it is possible to deduce a stronger version of Corollary 6.2.2. Suppose $\sum n_i\lambda[a_i(z)] \in A^*\otimes A^*$ is invariant under the switch map $a\otimes b\mapsto b\otimes a$. Then $\sum n_i D\big(a_i(z)\big) \equiv \text{const}$. This follows from $4(\lambda[z]+\lambda[z^{-1}]) = z^2\otimes z^2$ together with the existence of square roots in A and the polarization trick.

LECTURE 7

The Regulator Map. II

7.1. We have now a triangle

$$
\begin{array}{ccc}
K_3(\mathbb{C}) & \longrightarrow & K_2(R,I) \\
& {\scriptstyle D}\searrow \quad \swarrow{\scriptstyle D} & \\
& \mathbb{R} &
\end{array}
\tag{7.1.1}
$$

and a good function-theoretic hold on D. As applications we consider the case of a prime cyclotomic field $k = \mathbb{Q}(\zeta)$, $\zeta^\ell = 1$ and use D to obtain a regulator formula in the style of Kummer [**BS66**] for the value of the zeta function of k at $s = 2$. We also show how D leads to a measurable cohomology class in $H^3(\mathrm{SL}_2(\mathbb{C}), \mathbb{R})$ which is cohomologous to a multiple of the class arising from the volume form on the Poincaré upper half space $\mathrm{SL}_2(\mathbb{C})/\mathrm{SU}_2$.

7.2. Consider the commutative diagram

$$
\begin{array}{ccccccccc}
0 \to & \mathrm{tor}_1(\mathbb{C}^*,\mathbb{C}^*) & \to & (1+I)^* \otimes_{\mathbb{Z}} \mathbb{C}^* & \to & \coprod^{\circ}_{p\in\mathbb{C}^*} \mathbb{C}^* & \to & \mathbb{C}^* \otimes \mathbb{C}^* & \to 0 \\
& \downarrow & & \downarrow{\scriptstyle s} & & \| & & \downarrow & \\
0 \longrightarrow & K_3(\mathbb{C})^{\mathrm{ind}} & \longrightarrow & K_2(R,I) & \longrightarrow & \coprod^{\circ}_{p\in\mathbb{C}^*} \mathbb{C}^* & \to & K_2(\mathbb{C}) & \to 0
\end{array}
\tag{7.2.1}
$$

where the bottom line is (5.2.5) for $k = \mathbb{C}$, the top line arises from tensoring the sequence $0 \to (1+I)^* \to \coprod^{\circ}_{\mathbb{C}^*} \mathbb{Z} \to \mathbb{C}^* \to 0$ with $\mathbb{C}^*$, and the map s is induced from the product structure in K-theory, specifically from $K_1(R,I) \otimes K_1(\mathbb{C}) \to K_2(R,I)$. Taking the quotient yields the top line of

$$
\begin{array}{ccccccccc}
0 \longrightarrow & \frac{K_3(\mathbb{C})^{\mathrm{ind}}}{\mathrm{tor}_1(\mathbb{C}^*,\mathbb{C}^*)} & \longrightarrow & \frac{K_2(R,I)}{\{1+I,\mathbb{C}^*\}} & \xrightarrow{T} & \mathbb{C}^* \otimes_{\mathbb{Z}} \mathbb{C}^* & \longrightarrow & K_2(\mathbb{C}) & \longrightarrow 0 \\
& \uparrow{\scriptstyle \theta} & & \uparrow{\scriptstyle \eta} & & \| & & \| & \\
0 \longrightarrow & \mathcal{B}(\mathbb{C}) & \longrightarrow & \mathcal{A}(\mathbb{C}) & \xrightarrow{\lambda} & \mathbb{C}^* \otimes \mathbb{C}^* & \longrightarrow & K_2(\mathbb{C}) & \longrightarrow 0.
\end{array}
\tag{7.2.2}
$$

The bottom line is (6.1.2) with $F = \mathbb{C}$.

PROPOSITION 7.2.1. *For $x \in \mathbb{C} \setminus \{0,1\}$ define*

$$\eta[x] = 6\left\langle \frac{xt}{(t-1)^2 - xt^2}, \frac{t}{t-1} \right\rangle \in K_2(R, I).$$

With this η, the diagram (7.2.2) is commutative.

PROOF. $\mathbb{C}^* \subset K_0(\mathbb{P}^1_{\mathbb{C}}, \{0, \infty\})$ and the divisor class map

$$d\colon \coprod_{p\in\mathbb{C}^*}^{\circ} \mathbb{Z} \to \mathbb{C}^*$$

is given as folows: For δ a divisor find f on $\mathbb{P}^1$ such that $(f) = \delta$. Then $d(\delta) = f(0)f(\infty)^{-1}$. There is a commutative diagram

$$\tag{7.2.3} \begin{array}{ccc} K_2(R) & \xrightarrow{\substack{\text{tame}\\ \text{symbol}}} & \coprod^{\circ}_{p\in\mathbb{C}^*} \mathbb{C}^* \\ \uparrow & & \downarrow{\scriptstyle d\otimes \mathrm{Id}} \\ K_2(R, I) & \xrightarrow{T} & \mathbb{C}^* \otimes \mathbb{C}^*. \end{array}$$

Note

$$\tag{7.2.4} \eta[x] = \left\{1 - \frac{xt^2}{(t-1)^3 - xt^2(t-1)}, \frac{t}{t-1}\right\}^6$$

as a symbol. Since

$$\mathrm{tame}\{f, g\} = \sum_{y\in\mathbb{P}^1} (-1)^{\mathrm{ord}_y f\cdot \mathrm{ord}_y g} \frac{f(y)^{\mathrm{ord}_y g}}{g(y)^{\mathrm{ord}_y f}}$$

a straightforward calculation yields (writing $u = t/(t-1)$, $\zeta^3 = 1$)

$$\tag{7.2.5} \begin{aligned} \mathrm{tame}\left\{1 - \frac{xt^2}{(t-1)^3 - xt^2(t-1)}, \frac{t}{t-1}\right\}^6 &= \mathrm{tame}\left\{\frac{1-xu^3}{1-xu^2}, u\right\}^6 \\ = \big((-1)^6|_{u=\infty}\big) + \big(x^2|_{u=x^{-1/3}}\big) &+ \big(x^2|_{u=\zeta x^{-1/3}}\big) \\ + \big(x^2|_{u=\zeta^2 x^{-1/3}}\big) - \big(x^3|_{u=x^{-1/2}}\big) &- \big(x^3|_{u=-x^{-1/2}}\big). \end{aligned}$$

A function f with divisor in u coordinates

$$(f) = 2(x^{-1/3}) + 2(\zeta x^{-1/3}) + 2(\zeta^2 x^{-1/3}) - 3(x^{-1/2}) - 3(-x^{-1/2})$$

is

$$f = \frac{(1 - xu^3)^2}{(1 - xu^2)^3}.$$

Since $u(0) = 0$, $u(\infty) = 1$, we find $f(0)f(\infty)^{-1} = 1 - x$. It now follows from (7.2.3) that $T_\eta[x] = (1-x) \otimes x = \lambda[x]$ as claimed. Note that η determines a map

$$\theta\colon \mathcal{B}(\mathbb{C}) \to K_3(\mathbb{C})^{\mathrm{ind}}/\mathrm{tor}_1(\mathbb{C}^*, \mathbb{C}^*). \qquad \square$$

REMARK 7.2.2. The above discussion was completely algebraic, and would be valid with $\mathbb{C}$ replaced by any algebraically closed field.

EXAMPLE 7.2.3. Let $\zeta \in \mathbb{C}$ be a primitive nth root of 1 for some $n > 2$, and let $k = \mathbb{Q}(\zeta)$. Then $n[\zeta] \in \mathcal{B}(k)$. Also $\eta[\zeta] \in K_2(R_k, I_k)$, and it follows from a glance at the right hand side of (7.2.5) and the exact sequence

$$K_3(k) \to K_2(R_k, I_k) \to \coprod_{\substack{p \in \mathbb{P}^1_{k'} \\ p \neq 0, \infty}} k(p)^*$$

that $\theta(n[\zeta]) \in \mathrm{Image}\big(K_3(k) \to K_3(\mathbb{C})^{\mathrm{ind}}/\mathrm{tor}_1(\mathbb{C}^*, \mathbb{C}^*)\big)$. The kernel of this map can be shown to be torsion, so $\theta(n[\zeta]) \in K_3(k)/\text{torsion}$. We want to compute the image of this element under the regulator map D. For simplicity we restrict ourselves to the case $n = \ell$, an odd prime.

THEOREM 7.2.4. *With notation as above, the elements*

$$\theta(\ell[\zeta]), \theta(\ell[\zeta^2]), \ldots, \theta(\ell[\zeta^{(\ell-1)/2}])$$

form a basis for $K_3(k)_\mathbb{Q}$. Their image under the map

$$K_3(k)_\mathbb{Q} \to K_3(k \otimes_\mathbb{Q} \mathbb{R})_\mathbb{Q} \cong K_3(\mathbb{C})_\mathbb{Q}^{\oplus(\ell-1)/2} \xrightarrow{D^{\oplus(\ell-1)/2}} \mathbb{R}^{\oplus(\ell-1)/2}$$

generates a lattice of maximal rank whose volume is given by

$$2^{-(\ell-1)/2}\ell^{3(\ell-1)/4} \prod_{\chi \text{ odd}} |L(2,\chi)|$$

where χ runs through the odd characters of $(\mathbb{Z}/\ell\mathbb{Z})^*$, and $L(s,\chi) = \sum \chi(n)n^{-s}$ is the Dirichlet L-series.

Before proving Theorem 7.2.4, we verify a compatibility.

LEMMA 7.2.5. *The diagram*

$$\begin{array}{ccc} \mathcal{A}(\mathbb{C}) & \xrightarrow{\quad\eta\quad} & K_2(R, I) \\ & {\scriptstyle D}\searrow \quad \swarrow{\scriptstyle D} & \\ & \mathbb{R} & \end{array}$$

commutes, where the D on the left is the function D (Corollary 6.1.2) and that on the right is from (7.1.1).

PROOF. As a temporary expedient let D' denote the D on the right. With reference to (7.2.5) we find (replacing x by x^6)

$$D'\eta[x^6] = 6D\left(\left(\frac{1-x^6u^3}{1-x^6u^2}\right)^- * (u)\right)$$
$$= 6D\big((1-x^{-3}) + (1+x^{-3}) + (1) - (1-x^{-2}) - (1-\zeta x^{-2}) - (1-\zeta^2x^{-2})\big),$$

where $\zeta^3 = 1$.

Using $D(z^{-1}) = -D(z) = D(1-z)$, we reduce to proving

$$6D(x^3)+6D(-x^3)-6D(x^2)-6D(\zeta x^2)-6D(\zeta^2x^2)-D(x^6) = 0. \tag{7.2.6}$$

By Corollary 6.2.2 it suffices to show

$$6(x^3) + 6(-x^3) - 6(x^2) - 6(\zeta x^2) - 6(\zeta^2 x^2) - (x^6) \in \mathcal{B}(A).$$

(This will imply the left side of (7.2.6) is constant. Letting $x \to 0$ shows that constant is zero.) The requisite identity

$$[(1-x^3)\otimes x^{18}]\cdot[(1+x^3)\otimes x^{18}]\cdot[(1-x^2)\otimes x^{12}] \cdot[(1-\zeta x^2)\otimes x^{12}]\cdot[(1-\zeta^2x^2)\otimes x^{12}]\cdot[(1-x^6)\otimes x^6] = 1$$

is easily checked. □

7.3. We now turn to the proof of Theorem 7.2.4. Notice that Borel's theorem implies $K_3(k)_{\mathbb{Q}}$ has rank $(\ell-1)/2$, so it suffices to do the volume computation. (Note the L-series has a product expansion, so the volume $\neq 0$.) By (7.2.5)

$$D\eta[\zeta^i] = -\operatorname{Im}\int_0^{\zeta^i}\log(1-v)\frac{dv}{v} = \operatorname{Im}\left(\sum_{m=1}^{\infty}\frac{\zeta^{im}}{m^2}\right). \tag{7.3.1}$$

Write $D(i) = D\eta[\zeta^i]$. Note $D(-i) = -D(i)$, $D(0) = 0$, $D(i+\ell) = D(i)$. Thus we can view D as an odd function on $(\mathbb{Z}/\ell\mathbb{Z})^*$. Hence

$$D = \sum_{\substack{\chi \text{ odd character} \\ \text{of } (\mathbb{Z}/\ell\mathbb{Z})^*}} c_\chi \chi.$$

As representatives for the different equivalence classes of embeddings $k \hookrightarrow \mathbb{C}$ we can fix one $(\zeta \mapsto \zeta)$ and take as the others $\zeta \mapsto \zeta^j$, $2 \le j \le (\ell-1)/2$. The volume of the lattice spanned by the elements $\ell\eta[\zeta^i]$, $1 \le i \le (\ell-1)/2$, is then

$$\ell^{(\ell-1)/2}R, \quad R = \big|\det\big(D(ij)\big)\big|_{1\le i,j\le(\ell-1)/2}.$$

LEMMA 7.3.1.

$$R = \left(\frac{\ell-1}{2}\right)^{(\ell-1)/2} \prod_{\chi\ odd} |c_\chi|.$$

PROOF. Define $D_j(i) = D(ij)$ so $D_j = \sum_{\chi \text{ odd}} \chi(j) c_\chi \chi$. The matrix $D(ij)$ is the matrix of coefficients expressing the functions D_j in terms of the basis for the odd functions on $(\mathbb{Z}/\ell\mathbb{Z})^*$ given by

$$S_i(j) = \begin{cases} 0 & j \neq i, -i \\ 1 & j = i \\ -1 & j = -i \end{cases} \qquad 1 \le i \le \frac{\ell-1}{2}.$$

Since the odd χ form another basis for this space, R is necessarily a homogeneous function of degree $(\ell-1)/2$ in the c_χ. Clearly if some $c_\chi = 0$, $R = 0$ (the D_j won't span), so

$$R = |C| \prod |c_\chi|$$

for some universal constant C. To compute C, consider the function $F = \sum_{\chi \text{ odd}} \chi$. $F(ij) = \sum_\chi \chi(i)\chi(j)$. Ordering the χ, we can consider the matrix

$$M = \big(\chi(i)\big)_{\substack{\chi \text{ odd} \\ 1 \le i \le (\ell-1)/2}}.$$

Then

$$\big(F(ij)\big) = M^{\mathrm{t}} M, \quad |C| = |\det M|^2.$$

To compute $\det M$, let θ be a generator of the $(\ell-1)$st roots of 1, and let $g \in (\mathbb{Z}/\ell\mathbb{Z})^*$ be a generator. The numbers $1 \le i \le (\ell-1)/2$ form a set of coset representatives for $\{\pm 1\} \subset (\mathbb{Z}/\ell\mathbb{Z})^*$. Replacing this set by another in the definition of M changes $\det M$ by at most a sign, so we may consider instead $\det\big(\chi(i)\big)_{\chi \text{ odd}, i \in \mathcal{S}}$ where $\mathcal{S} = \{g^j \mid 1 \le j \le (\ell-1)/2\}$. We have

$$\chi(g^j) = \theta^{(2r+1)j} \quad \text{some } r = r(\chi), 0 \le r \le \frac{\ell-3}{2},$$

so the determinant becomes

$$\begin{aligned} \det\big(\theta^{(2r+1)j}\big)_{\substack{0 \le r \le (\ell-3)/2 \\ 1 \le j \le (\ell-1)/2}} &= \theta^{(\ell-1)(\ell+1)/8} \det\big(\theta^{2ri}\big)_{\substack{0 \le r \le (\ell-3)/2 \\ 1 \le i \le (\ell-1)/2}} \\ &= \pm \theta^{(\ell-1)(\ell+1)/8} \prod_{\substack{i,j=1 \\ i<j}}^{(\ell-1)/2} (\theta^{2j} - \theta^{2i}). \end{aligned}$$

Thus

$$\det M^2 = \pm\theta^{(\ell-1)(\ell+1)/4} \prod_{\substack{i,j=1 \\ i\neq j}}^{(\ell-1)/2} (\theta^{2j} - \theta^{2i}) = \pm \prod_{\substack{i,j=1 \\ i\neq j}}^{(\ell-1)/2} (1 - \theta^{2i-2j}).$$

Since $\prod_{k=1}^{(\ell-3)/2}(1-\theta^{2k}) = (\ell-1)/2$, we find

$$\det M^2 = \pm\left(\frac{\ell-1}{2}\right)^{(\ell-1)/2},$$

proving Lemma 7.3.1. □

LEMMA 7.3.2.

$$c_\chi = \frac{1}{\ell-1}\sum_{k=1}^{\ell-1}\overline{\chi}(k)D(k),$$

so

$$R = 2^{-(\ell-1)/2}\prod_{\chi\ odd}\left|\sum_{k=1}^{\ell-1}\overline{\chi}(k)D(k)\right|.$$

PROOF. This is simply the orthogonality relations for the χ. □

Recall now the formula for the Dirichlet L-series

$$L(s,\chi) = \frac{\tau(\chi)}{\ell}\sum_{k=1}^{\ell-1}\overline{\chi}(k)\sum_{n=1}^{\infty}\frac{\zeta^{-nk}}{n^s}$$

where $\zeta = e^{-2\pi i/\ell}$, $\tau(\chi) = \sum_{u \bmod \ell}\chi(u)\zeta^u$. If χ is odd we find using $|\tau(\chi)| = \ell^{1/2}$

$$|L(2,\chi)| = \ell^{-1/2}\left|\sum_{k=1}^{\ell-1}\overline{\chi}(k)D(k)\right|.$$

As a consequence

$$R = 2^{-(\ell-1)/2}\ell^{(\ell-1)/4}\prod_{\chi\text{ odd}}|L(2,\chi)|.$$

This completes the proof of Theorem 7.2.4. □

7.4. We now want to sketch a more geometric application of the function $D(x)$. These results will not be used in the sequel. Consider the *Poincaré upper half space* $\mathcal{H}$ as the interior of the ball, bounded by $S^2 = \mathbb{P}^1_{\mathbb{C}}$. Given four points in $\mathcal{H}$, it makes sense to talk about the geodesic tetrahedron they span (the metric being, of course, the Poincaré metric), and about the volume of this tetrahedron. There exist formulas for the volume [**Cox35**], but they are quite complicated.

Notice, however, that if one or more of the points "goes to ∞" (i.e., approaches a limiting position on $\mathbb{P}^1_{\mathbb{C}}$), the volume stays finite. Indeed, one can think of $\mathcal{H}$ as the upper half space given in x, y, z coordinates by $z > 0$, with the vertex p in question lying at ∞. In this case geodesics through p are vertical lines and the Poincaré metric looks like $(dx \wedge dy \wedge dz)/z^2$. Finiteness of volume follows from finiteness of the integral $\int_{z_0}^{\infty} dz/z^2$.

Suppose now that all four points lie at ∞. Given $g_1, g_2, g_3, g_4 \in \mathbb{P}^1_{\mathbb{C}}$, let $V(g_1, \ldots, g_4)$ denote the oriented volume of the tetrahedron they span.

LEMMA 7.4.1. *V satisfies the cocycle condition, given $g_1, \ldots, g_5 \in \mathbb{P}^1_{\mathbb{C}}$*

$$\sum_{i=1}^{5} (-1)^i V(g_1, \ldots, \hat{g}_i, \ldots, g_5) = 0. \tag{7.4.1}$$

PROOF. By taking the "geodesic span", the points $g_1, \ldots, g_5$ define a map $\Phi\colon \Delta^4 \to \overline{\mathcal{H}} = \mathcal{H} \cup \mathbb{P}^1_{\mathbb{C}}$, such that

$$\delta\Phi = \sum (-1)^i \Delta(g_1, \ldots, \hat{g}_i, \ldots, g_5)$$

where $\Delta(g_1, \ldots, \hat{g}_i, \ldots, g_5)$ denotes the geodesic tetrahedron with vertices $g_1, \ldots, \hat{g}_i, \ldots, g_5$. We can approximate Φ by maps $\Phi_\alpha\colon \Delta^4 \to \mathcal{H}$. Writing $\omega =$ volume form on $\mathcal{H}$, the left hand side of (7.4.1) is approximated by

$$\int_{\delta\Delta^4} \Phi_\alpha^* \omega = \int_{\Delta^4} \Phi_\alpha^*\, d\omega = 0. \qquad \square$$

THEOREM 7.4.2. *$V(z_1, \ldots, z_4) = \pm\frac{3}{2} D(\{z_1, \ldots, z_4\})$, where D is as in Corollary 6.1.2 and*

$$\{z_1, \ldots, z_4\} = \frac{(z_1 - z_2)(z_3 - z_4)}{(z_1 - z_4)(z_3 - z_2)}$$

is the cross-ratio.

LEMMA 7.4.3.

$$\sum_{i=1}^{5} (-1)^i D(\{z_1, \ldots, \hat{z}_i, \ldots, z_5\}) = 0.$$

PROOF. Notice that it suffices to show the left side is constant, because we can take, e.g., $z_1 = z_2$ and use the fact $D(0) = D(1) = D(\infty) = 0$ to deduce that the constant is zero.

To show that the desired expression is constant, we use (6.2.7). Notice

$$1 - \{z_1, \ldots, z_4\} = \{z_1, z_3, z_2, z_4\} = \frac{(z_1 - z_3)(z_2 - z_4)}{(z_1 - z_4)(z_3 - z_2)}.$$

Thus

$$\lambda\{z_1, \ldots, z_4\} = \big((z_1 - z_4)(z_3 - z_2) \otimes (z_1 - z_4)(z_3 - z_2)\big) \\ \cdot \prod\{\text{terms of the form } (z_i - z_j) \otimes (z_k - z_\ell)^{\pm 1}\}$$

where i, j, k, ℓ are *not* all distinct. The first term is symmetric and can be ignored. Consider a term $(z_i - z_j) \otimes (z_k - z_\ell)^{\pm 1}$. Since i, j, k, ℓ are not all distinct, there will be $m \in \{1, 2, 3, 4\}$, $m \neq i, j, k, \ell$. One checks then that the term $(z_i - z_j) \otimes (z_k - z_\ell)^{\mp 1}$ appears in the expansion for $(\lambda\{z_1, \ldots, \hat{z}_m, \ldots, z_5\})^{(-1)^m}$. It is occasionaly necessary to replace $(z_i - z_j) \otimes (z_k - z_\ell)$ by $(z_k - z_\ell) \otimes (z_i - z_j)^{-1}$, but this is permissible by symmetry. With these hints, the patient reader can check the details for himself. □

The key point is now

THEOREM 7.4.4. *The set of measurable functions*

$$f\colon \underbrace{\mathbb{P}^1_{\mathbb{C}} \times \cdots \times \mathbb{P}^1_{\mathbb{C}}}_{4\ \textit{times}} \to \mathbb{R}$$

which are invariant under the diagonal action of $\mathrm{PGL}_2(\mathbb{C})$ *and satisfy the cocycle condition* $\sum(-1)^i f(z_1, \ldots, \hat{z}_i, \ldots, z_5) = 0$ *forms a one dimensional* $\mathbb{R}$*-vector space generated by* $D(\{z_1, \ldots z_4\})$.

For the proof, let $G = \mathrm{PGL}_2(\mathbb{C})$ and define a complex of G-modules $C^\bullet$ by $C^i = \mathrm{Morph}\big((\mathbb{P}^1_{\mathbb{C}})^{i+1}, \mathbb{R}\big)$, $i \geq 0$, where "Morph" means measurable functions, identifying those which agree a.e. For details about properties of $C^\bullet$ see [**Moo76**]. Note in particular:

LEMMA 7.4.5. (i) *If* $U \subset (\mathbb{P}^1_{\mathbb{C}})^{i+1}$ *is open with complement of measure zero, then* $C^i \cong \mathrm{Morph}(U, \mathbb{R})$.

(ii) $C^\bullet$ *is a resolution of* $\mathbb{R}$ *by topological* G*-modules, so*

$$H^*_{\mathrm{cont}}(G, \mathbb{R}) \cong H^*(G, C^\bullet)$$

(hypercohomology computed as in [**Moo76**]*).*

(iii) $H^*(G, C^0) \cong H^*_{\mathrm{cont}}(B, \mathbb{R})$; $H^*(G, C^1) \cong H^*_{\mathrm{cont}}(T, \mathbb{R})$;

$$H^*(G, C^2) \cong \begin{cases} \mathbb{R} & * = 0 \\ 0 & * > 0 \end{cases}$$

where $T \subset B \subset G$ *are the torus and Borel subgroups.*

PROOF. (i) is clear, and (iii) follows from [**Moo76**, Theorem 6], together with (i). (Note $G/B \cong \mathbb{P}^1$ and $G/T \hookrightarrow \mathbb{P}^1 \times \mathbb{P}^1$ is open dense.) For (ii), the fact that functions in C^i are only defined a.e. makes it awkward to write down a contracting homotopy. Instead, let $f \in C^i$ be a cocycle for some $i \geq 1$:

$$\sum (-1)^j f(x_0, \ldots, \hat{x}_j, \ldots, x_{i+1}) = 0 \quad \text{(a.e.)},$$

and choose $y \in \mathbb{P}^1$ such that

$$f(x_0, \ldots, x_i) = \sum_{j=0}^{i} (-1)^j f(y, x_0, \ldots, \hat{x}_j, \ldots, x_i)$$

for almost all $(x_0, \ldots, x_i)$. Taking $g(x_0, \ldots, x_{i-1}) = f(y, x_0, \ldots, x_{i-1})$ we get $\delta g = f$. For $f \in C^0$, $\delta f = 0$ means $f(x_0) = f(x_1)$, i.e., $f \equiv \text{const}$. □

LEMMA 7.4.6. (i) $H^2_{\text{cont}}(B, \mathbb{R}) = (0)$;
(ii) $H^3_{\text{cont}}(G, \mathbb{R}) \cong \mathbb{R} \cong H^1_{\text{cont}}(T, \mathbb{R})$.

PROOF. We already did these computations in (3.3). □

LEMMA 7.4.7. (i) *The map*

$$\mathbb{R} \cong H^0(G, C^1) \xrightarrow{\delta} H^0(G, C^2) \cong \mathbb{R}$$

is an isomorphism.

(ii) *The map*

$$H^1_{\text{cont}}(B, \mathbb{R}) \cong H^1(G, C^0) \xrightarrow{\delta} H^1(G, C^1) \cong H^1_{\text{cont}}(T, \mathbb{R})$$

is twice the natural restriction map. In paricular, it is surjective.

PROOF. (i) A G-invariant map $f \colon \mathbb{P}^1 \times \mathbb{P}^1 \to \mathbb{R}$ is necessarily constant (a.e.), $f \equiv C$. Then $\delta f(x_0, x_1, x_2) = C - C + C = C$.

(ii) For $g \in G$, let $\bar{g} \in B\backslash G$, $\tilde{g} \in T\backslash G$ denote the images. The boundary map

$$\delta \colon \text{Morph}(B\backslash G, \mathbb{R}) \to \text{Morph}(T\backslash G, \mathbb{R})$$

is given by

$$(\delta f)(\tilde{g}) = p_0(f)(\tilde{g}) - p_\infty(f)(\tilde{g}),$$

where $p_j(f)(\tilde{g}) = f(j, \tilde{g})$, $j = 0, \infty$. Let $i \colon T \hookrightarrow B$ be the inclusion. The map p_0^* on cohomology coincides with i^*. Let $J = \left(\begin{smallmatrix} 0 & 1 \\ -1 & 0 \end{smallmatrix}\right) \in G$, so $\infty J = 0$, $0J = \infty$. Conjugation by J acts on $T \cong \mathbb{C}^*$ by $x \mapsto x^{-1}$, so it acts by the identity on $H^0(T, \mathbb{R})$ and by -1 on $H^1(T, \mathbb{R})$. Since p_∞^* is given in cohomology by $J \circ i^*$, the lemma follows. □

PROOF OF THEOREM 7.4.4. We compute with the spectral sequence

$$E_1^{p,q} = H^q(G, C^p) \implies H^{p+q}_{\text{cont}}(G, \mathbb{R}).$$

From the lemmas,

$$E_2^{1,1} = E_2^{0,2} = (0)$$
$$E_2^{3,0} \cong \operatorname{Ker}\left(\operatorname{Morph}_G\big((\mathbb{P}^1)^4, \mathbb{R}\big) \xrightarrow{\delta} \operatorname{Morph}_G\big((\mathbb{P}^1)^5, \mathbb{R}\big)\right)$$

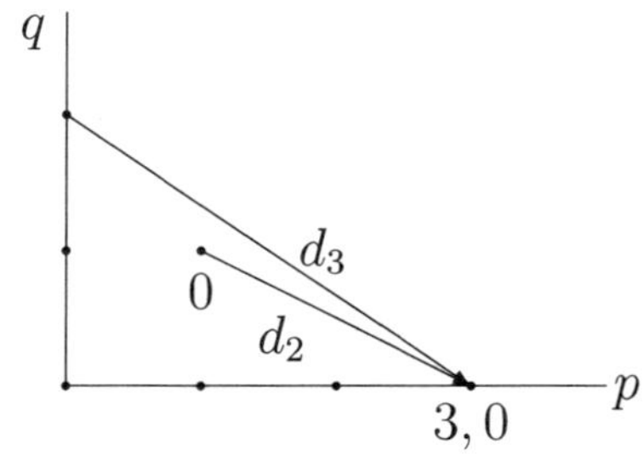

E_2

It follows that $E_2^{3,0} \cong E_\infty^{3,0} \hookrightarrow H^3_{\text{cont}}(G, \mathbb{R}) \cong \mathbb{R}$. Since $D(\{z_1, \ldots, z_4\}) \in E_2^{3,0}$ is non-trivial, we get $E_2^{3,0} \cong \mathbb{R}$, the statement of the theorem. $\square$

From Lemma 7.4.1 and Theorem 7.4.4 we conclude

$$V(z_1, \ldots, z_4) = CD(\{z_1, \ldots, z_4\})$$

for some constant C. Coxeter ([**Cox35**, p. 29]) computes the volume of the *regular* tetrahedron inscribed at ∞ to be

$$V = \frac{3\sqrt{3}}{4}\left(1 - \frac{1}{2^2} + \frac{1}{4^2} - \frac{1}{5^2} + \frac{1}{7^2} - \frac{1}{8^2} + \cdots\right).$$

The cross-ratio of the four vertices can be taken to be $e^{2\pi i/3} = \frac{1}{2}(-1 + \sqrt{-3})$ so

$$V = \pm C \cdot \operatorname{Im}\left(\sum_{n=1}^{\infty} \frac{e^{2\pi i n/3}}{n^2}\right) = \frac{\sqrt{3}}{2} C\left(1 - \frac{1}{2^2} + \frac{1}{4^2} - \frac{1}{5^2} + \cdots\right).$$

Thus $C = \frac{3}{2}$, proving Theorem 7.4.2.

REMARK 7.4.8. The analogy between $D(x)$ and $\log|x|$ comes out quite clearly. For one thing, viewing $\mathbb{P}^1_{\mathbb{C}} = \mathrm{GL}_2(\mathbb{C})/B$, $D \circ (\text{cross ratio})$ can be viewed as a homogeneous 3-cocycle just as $\log|\det|$ is a 1-cocycle. Also $\log|\text{cross ratio}|$ can be interpreted as the length of a geodesic. It would be very interesting to have a differential-geometric interpretation of the regulator on an elliptic curve discussed in the next section.

LECTURE 8

The Regulator Map and Elliptic Curves. I

8.1. Let E be an elliptic curve defined over $\mathbb{C}$. Our objective in the next few lectures will be to write down a regulator map

$$R\colon K_2(E) \to \mathbb{C}. \tag{8.1.1}$$

When E has complex multiplication over an imaginary quadratic field of class number 1, we will relate the zeta function, $\zeta(E/k, 2)$ ($k =$ Pray class field of the conductor of the Grössencharakter of E) to the value of R on certain elements in $K_2(E_k)$.

We fix $q = e^{2\pi i\tau} \in \mathbb{C}^*$ with $|q| < 1$ such that $E \cong \mathbb{C}^*/q^{\mathbb{Z}}$. Then

$$R = R_q = J_q + iD_q, \tag{8.1.2}$$

where J_q and D_q are real valued. As the notation indicates, R_q, J_q, D_q depend *a priori* on q. I have not yet worked out the modular behavior of these objects.

LEMMA 8.1.1. *The expression*

$$D_q(x) = \sum_{n\in\mathbb{Z}} D(xq^n)$$

defines a continuous function $D_q\colon E \to \mathbb{R}$.

PROOF. The fact that the expression in question converges uniformly as $n \to \infty$ follows from the description of $D(x)$ (Corollary 6.1.2). For $n \to -\infty$ we use Lemma 6.2.5 to write

$$D(xq^n) = -D(x^{-1}q^{-n}). \qquad \square$$

We will ultimately show

THEOREM 8.1.2. *D_q is a Steinberg function (cf. Definition 5.3.1) and hence induces a map* $D_q\colon K_2\big(\mathbb{C}(E)\big) \to \mathbb{R}$.

The imaginary part of our regulator is obtained by composing

$$K_2(E) \to K_2\big(\mathbb{C}(E)\big) \xrightarrow{D_q} \mathbb{R}.$$

To explain the real part J_q, we set

$$J(x) = \log|x| \log|1 - x|.$$

Let $\log|\ | \cdot \log|\ |\colon \coprod_{x\in\mathbb{C}^*} \mathbb{C}^* \to \mathbb{R}$ denote the map $f|_x \mapsto \log|f| \log|x|$.

LEMMA 8.1.3. *The diagram (notation as in Lecture 5) below commutes.*

$$\begin{array}{ccc} (1+I)^* \otimes_{\mathbb{Z}} \mathbb{C}(\mathbb{P}^1)^* & \xrightarrow{\text{tame}} & \coprod_{x\in\mathbb{C}^*} \mathbb{C}^* \\ \Big\downarrow{\scriptstyle (5.3.1)} & & \Big\downarrow{\scriptstyle \log|\ |\cdot\log|\ |} \\ \coprod_{x\in\mathbb{C}^*} \mathbb{Z} & \xrightarrow{J} & \mathbb{R} \end{array}$$

PROOF. Let $f = \prod(t-\alpha_j)^{d_j} \in (1+I)$, so $\sum d_j = 0$, $\prod \alpha_j^{d_j} = 1$, and let $g = c\prod(t-\beta_k)^{e_k} \in \mathbb{C}(\mathbb{P}^1)^*$. Starting with $f \otimes g$ in the upper left corner and going clockwise around the square, we obtain (with the natural convention that $\log|0|\log|1| = 0$)

$$\sum_{j,k} -d_j \log|\alpha_j| \cdot \log\left|c\prod_\ell(\alpha_j-\beta_\ell)^{e_\ell}\right| + e_k \log|\beta_k|\cdot\log\left|\prod_m(\beta_k-\alpha_m)^{d_m}\right|$$
$$= \sum_{j,k} d_j e_k \log|\beta_k\alpha_j^{-1}|\log|1-\beta_k\alpha_j^{-1}| + e\sum_j d_j(\log|\alpha_j|)^2$$

where $e = -\sum_k e_k = \operatorname{ord}_\infty g$. Since

$$(\log|\alpha|)^2 = \log|\alpha|\log|1-\alpha| + \log|\alpha^{-1}|\log|1-\alpha^{-1}|$$

the above expression coincides with $J\big((f)^- * (g)\big)$. □

It follows from the lemma that J is a relative Steinberg function, so we get

$$J\colon K_2(R,I) \to \mathbb{R}. \tag{8.1.3}$$

Notice that J factors through the tame symbol, so $J\big(K_3(\mathbb{C})\big) = 0$. Define

$$J_q(x) = \sum_{n=0}^{\infty} J(xq^n) - \sum_{n=1}^{\infty} J(x^{-1}q^n). \tag{8.1.4}$$

$J_q(x)$ is clearly continuous on $\mathbb{C}^*$, and we have

$$J_q(qx) - J_q(x) = -J(x^{-1}) - J(x) = -(\log|x|)^2. \tag{8.1.5}$$

LEMMA 8.1.4. *Let $F = \sum d_j(\alpha_j)$, $G = \sum e_k(\beta_k)$ be divisors on $\mathbb{C}^*$, and assume $\sum d_j = \sum e_k = 0$, $\prod \alpha_j^{d_j} = \prod \beta_k^{e_k} = 1$. Then F and G project to divisors of elliptic functions f, g on $E = \mathbb{C}^*/q^{\mathbb{Z}}$, and the expression*

$$J_q(F^- * G) = \sum_{j,k} d_j e_k J_q(\alpha_j^{-1}\beta_k)$$

depends only on the divisors of f and g.

PROOF. The first assertion is well-known. For the second, it suffices to consider

$$J_q(F'^- * G), \quad F' = \sum d_j(\alpha_j q^{n_j}), \sum d_j n_j = 0.$$

Write

$$J_{q,G}(x) = \sum_k e_k J_q(x\beta_k).$$

We have from (8.1.5)

$$J_{q,G}(qx) - J_{q,G}(x) = \sum e_k(\log|\beta_k x|)^2 = \sum e_k(\log|\beta_k|)^2$$

$$J_{q,G}(q^n x) - J_{q,G}(x) = \sum n e_k(\log|\beta_k|)^2.$$

$$J_q(F'^- * G) - J_q(F^- * G) = \sum_j d_j\big(J_{q,G}(\alpha_j^{-1}q^{-n_j}) - J_{q,G}(\alpha_j^{-1})\big)$$

$$= \sum_{j,k} -d_j n_j e_k(\log|\beta_k|)^2 = 0. \qquad \square$$

It follows immediately from (8.1.4) that the prescription

$$f \otimes g \mapsto J_q(F^- * G),$$

where F lifts (f) and G lifts (g), defines a homomorphism

(8.1.6) $$\mathbb{C}(E)^* \otimes_{\mathbb{Z}} \mathbb{C}(E)^* \xrightarrow{J_q} \mathbb{R}.$$

We will eventually show

THEOREM 8.1.5. *$J_q\big(f \otimes (1-f)\big) = 0$ for $f \not\equiv 1$, so J_q induces a map*

$$J_q\colon K_2(E) \to K_2\big(\mathbb{C}(E)\big) \to \mathbb{R}.$$

8.2. The proof of Theorems 8.1.2 and 8.1.5 require rather detailed knowledge of the behavior of functions like

(8.2.1) $$F(w) = \prod_{\substack{n=0\\ j}}^{\infty}(1-\alpha_j q^n w^{-1})^{d_j} \prod_{\substack{n=1\\ j}}^{\infty}(1-\alpha_j^{-1}q^n w)^{d_j}$$

where $\sum d_j = 0$, $\prod \alpha_j^{d_j} = 1$. Note

$$F(q^{-1}w)F(w)^{-1} = \prod_j(1-\alpha_j^{-1}w)^{d_j} \cdot \prod_j(1-\alpha_j w^{-1})^{-d_j}$$

$$= \prod_j(-\alpha_j w)^{-d_j} = 1,$$

so $F(w)$ is the pullback to $\mathbb{C}^*$ of an elliptic function with divisor $\sum d_j(\alpha_j)(\mathrm{mod} q^{\mathbb{Z}})$. We fix a constant $K \neq 0$, 1, and for $N > 0$ we let A_N denote the annulus

$$|q|^N \leq |z| \leq |q|^{-N}. \tag{8.2.2}$$

Also for $N \geq 0$ an integer we write

$$\begin{gathered} F_0(w) = \prod_j (w - \alpha_j)^{d_j}, \quad F_N(w) = \prod_{|r| \leq N} F_0(q^r w) \\ T_N(w) = \prod_{\substack{n \geq N+1 \\ j}} (1 - \alpha_j q^n w^{-1})^{d_j} (1 - \alpha_j^{-1} q^n w)^{d_j} \end{gathered} \tag{8.2.3}$$

so $F_N(w) T_N(w) = F(w)$.

LEMMA 8.2.1. *There exists an $R > 0$ independent of N such that $F_N(w) - K$ has no zeros off the annulus A_{N+R}.*

PROOF. Replacing $F(w)$ by $F(w^{-1})$ it will suffice to show $F_N(w) - K$ has no zero inside a circle of radius $|q|^{N+R}$. Then replacing w by $q^N w$ it suffices to show $F_N(q^N w) - K$ has no zero inside a circle of radius δ for some fixed δ.

Consider the power series expansion

$$F_0(w) = 1 + a_1 w + a_2 w^2 + \cdots$$

and choose a constant $s > 0$ such that $|a_n| \leq s^n$, all n. We have

$$F_N(q^N w) = \prod_{r=0}^{2N} F_0(q^r w) \ll \prod_{r=0}^{2N} \frac{1}{(1 - |q|^r sw)} \ll \prod_{r=0}^{\infty} \frac{1}{(1 - |q|^r sw)}$$

where $\ll$ means term by term domination of the coefficients of the power series expansions. Since the coefficients on the right are all positive real, we find

$$|F_N(q^N w) - K| \geq |K - 1| - \left| 1 - \prod_0^{\infty} \frac{1}{(1 - |q|^r s|w|)} \right| > 0$$

for $|w| \leq \delta$, some δ. □

In what follows, N and ϵ are parameters with N large and ϵ small. The notation

$$\psi(N, \epsilon) = O(N\epsilon) + O(1)$$

will mean there exist constants C, C' independent of N, ϵ such that

$$\psi(N, \epsilon) \leq CN\epsilon + C'.$$

The notation $Z(g, R)$ will denote the number of zeros of a function $g(w)$ on A_R.

LEMMA 8.2.2. *Given* $0 < \epsilon < 1$ *irrational,* $R > 0$*, and* $N_0 > 0$*, there exists* $N \geq N_0$ *such that*

$$Z(F_N(w) - K, N(1-\epsilon) + R) = Z(F(w) - K, N(1-\epsilon) + R).$$

In particular, the number of zeros of $F_N(w) - K$ *off* $A_{N(1-\epsilon)+R}$ *is* $O(N\epsilon) + O(1)$ *for some sequence of* N*'s* $\to \infty$*, where the sequence of* N*'s may depend on* ϵ*.*

PROOF. The second statement follows from the first because if $F_0(w)$ has (say) d poles, then $F(w) - K$ will have $dN(1-\epsilon) + O(1)$ poles on $A_{N(1-\epsilon)+R}$ and (up to $O(1)$) the same number of zeros. Since $F_N(w) - K$ has Nd poles and Nd zeros, it will have (by the first part of the lemma) $Nd\epsilon + O(1)$ zeros off $A_{N(1-\epsilon)+R}$.

To prove the lemma, it suffices to find $N \geq N_0$ such that

$$|F_N(w) - F(w)| = |F_N(w)||T_N(w) - 1| < |F(w) - K|$$

at the boundary $\delta A_{N(1-\epsilon)+R}$. In fact, this will insure that

$$\left|\frac{F_N(w) - K}{F(w) - K} - 1\right| < 1 \text{ on } \delta A_{N(1-\epsilon)+R}$$

whence

$$\int_{\delta A_{N(1-\epsilon)+R}} \operatorname{dlog}\left(\frac{F_N(w) - K}{F(w) - K}\right) = 0.$$

Since the numbers of poles of $F_N(w) - K$ and $F(w) - K$ on $A_{N(1-\epsilon)+R}$ coincide, we must have

$$Z(F(w) - K, N(1-\epsilon) + R) = Z(F_N(w) - K, N(1-\epsilon) + R)$$

as claimed.

Write

$$\begin{gathered} F(w) - K = \mu \prod_{\substack{n=0 \\ k}}^{\infty} (1 - \beta_k q^n w^{-1})^{e_k} \prod_{\substack{n=1 \\ k}}^{\infty} (1 - \beta_k^{-1} q^n w)^{e_k} \\ \sum e_k = 0, \quad \prod \beta_k^{e_k} = 1, \quad \mu \in \mathbb{C}. \end{gathered} \tag{8.2.4}$$

Fix some compact region $V \subset \mathbb{R}/\mathbb{Z}$ such that $V = -V \neq \varnothing$ and such that V does not contain any of the points

$$\frac{\log|\beta_k|}{\log|q|}, \quad \frac{\log|\alpha_j|}{\log|q|} \quad (\text{mod } 1).$$

Take $N \geq N_0$ such that $N\epsilon + R \pmod 1 \in V$. (This is possible because ϵ is irrational.) Then I claim

(i) $|F(w) - K| \geq \delta = \delta(V) > 0$ on $\delta A_{N(1-\epsilon)+R}$
(ii) $|F_N(w)| = O(1)$ on $\delta A_{N(1-\epsilon)+R}$

(iii) $|T_N(w) - 1| = O(|q|^{N\epsilon})$ on $A_{N(1-\epsilon)+R}$

where δ in (i) and the "big O" in (ii), (iii) are independent of N. Indeed, (i) follows because $N\epsilon \equiv V$ implies $\beta_k q^n w^{-1} \neq 1$ on $\delta A_{N(1-\epsilon)+R}$ so $F(w) - K$ has no zeros on $\delta A_{N(1-\epsilon)+R}$. The fact that δ can be chosen independent of N comes from the fact that $F(qw) = F(w)$.

For (ii), we may replace $F_N(w)$ by $F_N(w^{-1})$ so it suffices to show boundedness on the inner contour Γ of $\delta A_{N(1-\epsilon)+R}$. Consider the set S of all integers n such that there exists a j with $|\alpha_j q^n w^{-1}| \leq 2$ *and* a j' with $|\alpha_{j'} q^n w^{-1}| \geq \frac{1}{2}$ on Γ. The number of elements in S can be bounded independently of N. Since $|1 - \alpha_j q^n w^{-1}|$ is bounded below on Γ,

$$\prod_{\substack{n \in S \\ j}} (1 - \alpha_j q^n w^{-1})^{d_j} (1 - \alpha_j^{-1} q^n w)^{d_j}$$

is bounded on Γ. Suppose $n \notin S$ and (say) $|\alpha_j q^n w^{-1}| \geq 2$ for all j. Dividing through by $\prod (\alpha_j q^n w^{-1})^{d_j} = 1$ we find

$$\prod_j (1 - \alpha_j q^n w^{-1})^{d_j} = \prod_j (\alpha_j^{-1} q^{-n} w - 1)^{d_j}.$$

In this way terms for $n \notin S$ can be majorized by products

$$\prod_n (1 - |q|^n \cdot \text{const})$$

which are bounded independent of N.

The bound $|T_N(w) - 1| = O(|q|^{N\epsilon})$ on $A_{N(1-\epsilon)+R}$ is immediate. The lemma is proved by combining (i), (ii), and (iii). □

LEMMA 8.2.3. *Let $F(w) - K$ be given by (8.2.4), and let ϵ be sufficiently small so the circle γ_k of radius ϵ about β_k does not contain 0 or any other $\beta_{k'}$. Assume also (for simplicity) that all $e_k = \pm 1$. Then for $N \gg 0$, any zero or pole of $F_N(w) - K$ or $A_{N(1-\epsilon)+R}$ lies inside exactly one translate $\gamma_k \cdot q^r$. No two roots lie in the same translate.*

PROOF. Take $N \gg 0$, so that

$$|K||T_N(\beta) - 1| < \inf_{w \in \gamma_k} |K - F(w)|$$

for all $\beta \in A_{N(1-\epsilon)+R}$. Notice that $K - F_N(w)$, $F_N(w)$, $F(w)$, and $K - F(w)$ all have the same poles on $A_{N(1-\epsilon)+R}$ so we need only be concerned with zeros. If $\beta \in A_{N(1-\epsilon)+R}$ is a zero of $F_N(w) - K$

$$0 = (F_N(\beta) - K)T_N(\beta) = \big(1 - T_N(\beta)\big)K + (F(\beta) - K)$$

so

$$|F(\beta) - K| < \inf_{w \in \gamma_j} |F(w) - K| = \inf_{w \in \gamma_j q^r} |F(w) - K|.$$

Since all the zeros of $F(w) - K$ lie inside the $\gamma_k q^r$ and since $e_k = \pm 1$ it follows from Rouché's Theorem that β lies in a unique $\gamma_j q^r$. $\square$

Given ϵ small and irrational, we fix now $N \gg 0$ such that Lemmas 8.2.2 and 8.2.3 hold, where $R > 0$ is fixed such that $\gamma_k q^r \subset A_{N(1-\epsilon)+R}$ for $|r| \leq N(1-\epsilon)$. Let $\beta_k(N,r)$ denote the unique singularity of $F_N(w) - K$ inside $\gamma_k q^r$, $|r| \leq N(1-\epsilon)$. Let $\{\beta'_N\}$ denote the other zeros and poles of $F_N(w) - K$. Let $e_{\beta'_N}$ = multiplicity of β'_N.

LEMMA 8.2.4. *$\sum e_{\beta'_N} = 0$ and $\sum e_{\beta'_N} \log|\beta'_N| = O(N\epsilon) + O(1)$ as $N \to \infty$.*

PROOF. We know $\sum e_k = 0$ because $F(w) - K$ is elliptic, and also $\sum e_{\beta_N} = 0$ where β_N runs through all singularities of $F_N(w) - K$ (because $K \neq 1 \implies 0$ and ∞ are not singularities of $F_N(w) - K$.) It follows that $\sum e_{\beta'_N} = 0$ as well. Also

$$\sum e_{\beta_N} \log|\beta_N| = \log|F_N(0) - K| = \log|1 - K|.$$

Since

$$\big|\log|\beta_k(N,r)| - \log|\beta_k q^r|\big| = O(\epsilon)$$
$$\sum_k e_k \log|\beta_k q^r| = 0,$$

and there are $O(N)$ $\beta_k(N,r)$'s, we may drop from the sum the β_N's corresponding to $\beta_k(N,r)$'s and get

$$\sum e_{\beta'_N} \log|\beta'_N| = O(N\epsilon) + O(1).$$

$\square$

LECTURE 9

The Regulator Map and Elliptic Curves. II

9.1. Our first objective is

THEOREM 9.1.1. *For any elliptic function f and any constant K, $J_q(f \otimes (K-f)) = 0$ (notation as in (8.1.6)).*

PROOF. From the description of J_q, it is clear that $J_q(L \otimes g) = 0$ for any constant L and any elliptic function g. We may therefore replace f by Lf and assume the induced function f on $\mathbb{C}^*$ looks like

$$F(w) = \prod_j \left(\prod_{n=0}^{\infty} (1 - \alpha_j^{-1} q^n w)^{d_j} \prod_{n=1}^{\infty} (1 - \alpha_j q^n w^{-1})^{d_j} \right).$$

Further, since the expression $J_q(f \otimes (K-f))$ is continuous as a function of the divisors (f) and $(K-f)$, we may assume $K \neq 0$, 1, and f is general, so

$$F(w) - K = \mu \prod_k \left(\prod_{n=0}^{\infty} (1 - \beta_k^{-1} q^n w)^{e_k} \prod_{n=1}^{\infty} (1 - \beta_k q^n w^{-1})^{e_k} \right)$$

with $e_k = \pm 1$.

Let the map $(1+I)^* \otimes \mathbb{C}(\mathbb{P}^1)^* \to \mathbb{R}$ in Lemma 8.1.3 be denoted by $f \otimes g \mapsto J(f \otimes g)$. It is an easy exercise from the diagram in Lemma 8.1.3 to show $J(f \otimes (K-f)) = 0$ for $f \in (1+I)^*$, $K \in \mathbb{C}^*$. Write $J_f(x) = \sum d_j J(\alpha_j^{-1} x)$, where $(f) = \sum d_j(\alpha_j)$. (This notation is slightly different from that in the proof of (8.1.4).) If $(K-f) = \sum e_k(\beta_k)$ and $K-f$ has no zero or pole at 0 or ∞, we find

$$\sum e_k J_f(\beta_k) = 0.$$

Apply this with $f = F_N = \prod_{|r| \le N} F_0(q^r w)$, $F_0 = \prod_j (w - \alpha_j)^{d_j}$. With notation as in Lecture 8, we find

$$\sum_{\substack{|r| \le N(1-\epsilon) \\ k}} e_k J_{F_N}\big(\beta_k(N,r)\big) + \sum e_{\beta'_N} J_{F_N}(\beta'_N) = 0. \tag{9.1.1}$$

LEMMA 9.1.2.

$$\sum e_{\beta'_N} J_{F_N}(\beta'_N) = O(N\epsilon) + O(1).$$

PROOF. Let $C = C_F = \sum_j d_j(\log|\alpha_j|)^2$. By (8.2.4) it suffices to show

$$\sum_{\beta'_N} e_{\beta'_N}\left(J_{F_N}(\beta'_N) + C\Big(\frac{\log|\beta'_N|}{\log|q|} - N\Big)\right) = O(N\epsilon) + O(1).$$

Dropping the sum, this will follow if we know for $\beta = \beta'_N$

$$J_{F_N}(\beta) + \left(\frac{\log|\beta|}{\log|q|} - N\right)C = O(1). \tag{9.1.2}$$

Write $F_0^-(w) = \prod_j (w - \alpha_j^{-1})^{d_j}$, and note

$$J_{F_0}(x) + J_{F_0^-}(x^{-1}) = C.$$

It follows from Lemma 8.2.1 that

$$\left|\max\left(-N, \frac{-\log|\beta|}{\log|q|}\right) + \frac{\log|\beta|}{\log|q|}\right| = O(1),$$

so we get

$$\begin{aligned} J_{F_N}(\beta) &= \sum_{|r|\le N} J_{F_0}(\beta q^r) \\ &= \sum_{r=\max(-N,-\log|\beta|/\log|q|)}^{N} J_{F_0}(\beta q^r) - \sum_{r=-N}^{\max(-N,-\log|\beta|/\log|q|)} J_{F_0^-}(\beta^{-1}q^{-r}) \\ &\qquad + \left(N - \frac{\log|\beta|}{\log|q|}\right)C + O(1). \end{aligned}$$

Note that in the ranges indicated for the summations, $|\beta q^r|$ and $|\beta^{-1}q^{-r}| < 1$. For $|\beta q^{N_0}| < 1$, however, one easily checks that $\sum_{n=N_0}^{\infty} |J(\beta q^n)| = O(1)$ uniformly in β. Thus the two sums on the right are $O(1)$, and (9.1.2) follows. □

We write

$$J_{F,q}(x) = \sum d_j J_q(\alpha_j^{-1}x)$$

(again note that α_j^{-1}; a change from the notation in the proof of Lemma 8.1.4).

LEMMA 9.1.3. $|J_{F,q}(x) - J_{F_N}(x) + NC_F| = O(N\epsilon|q|^{N\epsilon})$ *as* $N \to \infty$, *uniformly in* x *for* $x \in A_{N(1-\epsilon)+R}$.

PROOF. We have (because $J(x)+J(x^{-1})=(\log|x|)^2$):

$$\begin{aligned} J_{F_N}(x) &= \sum_{\substack{|r|\le N\\ j}} d_j J(\alpha_j^{-1}q^r x) \\ &= \sum_{\substack{r=0\\ j}}^{N} d_j J(\alpha_j^{-1}q^r x) - \sum_{\substack{r=1\\ j}}^{N} d_j J(\alpha_j q^r x^{-1}) + \sum_{\substack{r=1\\ j}}^{N} d_j (\log|\alpha_j^{-1}q^{-r}x|)^2, \end{aligned}$$

so

$$|J_{F,Q}(x) - J_{F_N}(x) + NC_F| \le \left|\sum_{\substack{r=N+1\\ j}}^{\infty} d_j J(\alpha_j^{-1}q^r x)\right| + \left|\sum_{\substack{r=N+1\\ j}}^{\infty} d_j J(\alpha_j q^r x^{-1})\right|.$$

Lemma 9.1.3 now follows from

SUBLEMMA 9.1.4. *If* $|xq^{N_0}| \le M$, *then* $\sum_{n=N_0}^{\infty} |J(xq^n)| = O(M\log M)$ *as* $M \to 0$.

PROOF. $|J(xq^n)| = O\big(M|q|^{n-N_0}\log(M|q|^{n-N_0})\big)$. Since the series

$$\sum_{n=0}^{\infty} |q|^n \frac{\log(M|q|^n)}{\log M}$$

can be bounded uniformly in M for $M \ll 1$, we get

$$\sum |J(xq^n)| = O(M\log M).$$

This proves Sublemma 9.1.4 and also Lemma 9.1.3. □

Returning now to the proof of Theorem 9.1.1, we can rewrite (9.1.1) (since $\sum e_k = 0$):

$$\sum_{\substack{|r|\le N(1-\epsilon)\\ k}} e_k J_{F,q}\big(\beta(N,r)\big) = O(N\epsilon) + O(1) + O(N^2\epsilon|q|^{N\epsilon}). \tag{9.1.3}$$

LEMMA 9.1.5. *For* $x,\ x' \in \mathbb{C}^*$, $|x-x'| \le \epsilon|x'|$, *we have*

$$|J_{F,q}(x) - J_{F,q}(x')| = O(\epsilon\log\epsilon), \quad \epsilon \to 0$$

uniformly in x, x'.

PROOF. We have $J_{F,q}(qx) - J_{F,q}(x) = -C_F$, so

$$J_{F,q}(qx) - J_{F,q}(qx') = J_{F,q}(x) - J_{F,q}(x').$$

We may assume, therefore, that $x' \in A_1$. Note

$$\begin{aligned} J(xq^r) - J(x'q^r) &= \log|xq^r|\log|1-xq^r| - \log|x'q^r|\log|1-x'q^r| \\ &= \log|xq^r|\log\left|\frac{1-xq^r}{1-x'q^r}\right| + \log\left|\frac{x}{x'}\right|\log|1-x'q^r|. \end{aligned}$$

Since

$$\frac{1 - xq^r}{1 - x'q^r} = 1 + \frac{q^r(x' - x)}{1 - x'q^r}$$

and $|1 - x/x'| \le \epsilon$, we see

$$|J(xq^r) - J(x'q^r)| = O(\epsilon |q|^r r) \quad \text{as } r \to \infty$$

uniformly for $x' \in A_1$.

It suffices to show, therefore, that

$$|J(x) - J(x')| = O(|x - x'| \log |x - x'|)$$

on some fixed disk $B_a(0)$ of radius a about 0. Since $J(x) = J(1 - x)$ we may assume x, $x' \notin B_{1/3}(1)$, and since the inequality is more or less trivial on any compact region where J has continuous partial derivatives, we may assume a small, say $a \le e^{-1}/4$.

Suppose first (say) $|x| \le 2|x - x'|$. Then

$$|x'| \le |x - x'| + |x| \le 3|x - x'|.$$

Since $\log |1 - x| = O(|x|)$ on $B_a(0)$, and since the function $y|\log y|$ is monotone increasing for $y \le e^{-1}$, we get $J(x)$, $J(x')$ individually $O(|x - x'| \log |x - x'|)$.

The remaining case to consider is when $|x|$, $|x'| > 2|x - x'|$, so $\frac{1}{2} > |1 - x/x'|$ and $\log |x/x'| = O(|1 - x/x'|)$. We have

$$\begin{aligned}
&\big|\log |x| \log |1 - x| - \log |x'| \log |1 - x'|\big| \\
&\quad \le \big|\log |x|\big| \, \big|\log |1 - x| - \log |1 - x'|\big| + \left|\log \left|\frac{x}{x'}\right|\right| \, \big|\log |1 - x'|\big| \\
&\quad = O(|x - x'| \log |x - x'|) + O\left(\left|1 - \frac{x}{x'}\right| |x'|\right) \\
&\quad = O(|x - x'| \log |x - x'|).
\end{aligned}$$

□

We can now finish off the proof of Theorem 9.1.1. Equation (9.1.3) can be rewritten

$$\sum_{\substack{|r| \le N(1-\epsilon) \\ k}} e_k J_{F,q}(\beta_k q^r) = O(N\epsilon) + O(1) + O(N^2 \epsilon |q|^{N\epsilon}) + O(N\epsilon \log \epsilon)$$

Since $J_{F,q}(qx) - J_{F,q}(x) = -C_F$ and $\sum e_k = 0$, this becomes

$$\begin{aligned}
\big(1 + 2N(1 - \epsilon)\big) \sum_k e_k J_{F,q}(\beta_k) \\
= O(N\epsilon) + O(1) + O(N^2 \epsilon |q|^{N\epsilon}) + O(N\epsilon \log \epsilon).
\end{aligned}$$

Dividing by $2N(1 - \epsilon) + 1$ and letting $N \to \infty$ gives

$$\sum_k e_k J_{F,q}(\beta_k) = O(\epsilon \log \epsilon).$$

Now let $\epsilon \to 0$. □

9.2. The proof of the corresponding result for D_q is similar but slightly less complicated.

THEOREM 9.2.1. *D_q is a Steinberg function (Definition 5.3.1) on E.*

PROOF. We keep the same notations (F, F_N, $F-K$, $\beta_k(N,r)$, etc.) as above. We have, since D is a relative Steinberg function

$$0 = \sum_{\substack{|r|\le N(1-\epsilon)\\ k}} e_k D_{F_N}(\beta_k(N,r)) + \sum e_{\beta'_N} D_{F_N}(\beta'). \tag{9.2.1}$$

Note $\sum_{n\in\mathbb{Z}} |D(xq^n)|$ converges, so $|D_{F_N}(x)| = O(1)$ uniformly in x and N. Since the number of β'_N is $O(N\epsilon) + O(1)$, we get

$$\sum_{\substack{|r|\le N(1-\epsilon)\\ k}} e_k D_{F_N}(\beta_k(N,r)) = O(N\epsilon) + O(1). \tag{9.2.2}$$

LEMMA 9.2.2. *$|D_{F_N}(x) - D_{F,q}(x)| = O(N\epsilon|q|^{N\epsilon})$ as $N \to \infty$ uniformly in x for $x \in A_{N(1-\epsilon)+R}$.*

PROOF. It suffices to show

$$\left|\sum_{n=N+1}^{\infty} D(xq^n)\right| = O(N\epsilon|q|^{N\epsilon})$$

on $A_{N(1-\epsilon)+R}$. This is an easy consequence (cf. the proof of Sublemma 9.1.4) of the obvious bound

$$D(y) = O(|y|\log|y|), \quad |y| \to 0.$$ □

Substituting in (9.2.2),

$$\sum_{\substack{|r|\le N(1-\epsilon)\\ k}} e_k D_{F,q}(\beta_k(N,r)) = O(N\epsilon) + O(1) + O(N^2\epsilon|q|^{N\epsilon}). \tag{9.2.3}$$

LEMMA 9.2.3. *For x, $x' \in \mathbb{C}^*$, $|x-x'| \le \epsilon|x'|$ we have*

$$|D_{F,q}(x) - D_{F,q}(x')| = O(\epsilon\log\epsilon), \quad \epsilon \to 0$$

uniformly in x, x'.

PROOF. It suffices to show

$$|D(xq^r) - D(x'q^r)| = O\big(|x-x'|\big(|\log|x-x'|| + |r|\big)|q|^{|r|}\big)$$

uniformly for x in an annulus A. For this note $|x-x'| = O(|1/x-1/x'|)$ on A, and $D(x^{-1}) = -D(x)$ so we can replace xq^r by $x^{-1}q^{-r}$ and assume $r \geq 0$. It thus suffices to show

$$|D(x) - D(x')| = O(|x - x'| \log |x - x'|)$$

for x, $x' \in B_a(0)$, the disk of radius a about 0. Since $D(x) = -D(1-x)$, we may assume x, $x' \notin B_{1/3}(1)$, and since further the estimate is easy on compact sets on which D has continuous partial derivatives, we may assume $a \leq e^{-1}/4$. The argument now is identical with the proof of Lemma 9.1.5. □

We can now rewrite (9.2.3)

$$\sum_{\substack{|r|\leq N(1-\epsilon) \\ k}} e_k D_{F,q}(\beta_k q^r) = O(N\epsilon) + O(1) + O(N^2\epsilon |q|^{N\epsilon}) + O(N\epsilon \log \epsilon).$$

Divide by $2N(1-\epsilon)+1$ and let $N \to \infty$, getting

$$\sum_k e_k D_{F,q}(\beta_k) = O(\epsilon \log \epsilon).$$

Now let $\epsilon \to 0$. □

LECTURE 10

Elements in $K_2(E)$ of an Elliptic Curve E

10.1. We now have a regulator map

$$R_q\colon K_2\big(\mathbb{C}(E)\big) \to \mathbb{C}$$

which induces by restriction a regulator on the global $K_2(E)$. Abusing notation, we will also write $R_q = J_q + iD_q$ for the function on $\mathbb{C}^*$. Thus, if f, g are elliptic functions, $\sum d_j = \sum e_k = 0$, $\prod \alpha_j^{d_j} = \prod \beta_k^{e_k} = 1$, and

$$(\tilde{f}) = \sum d_j(\alpha_j), \quad (\tilde{g}) = \sum e_k(\beta_k)$$

are liftings of (f), (g) to $\mathbb{C}^*$, we have

$$R_q\{f, g\} = \sum d_j e_k R_q(\alpha_j^{-1}\beta_k).$$

The next step will be to construct interesting elements in $K_2(E)$ to which we can apply R_q. Recall the exact sequence (valid for any smooth curve over a field k)

$$K_2(E) \to K_2\big(k(E)\big) \xrightarrow{\text{tame}} \coprod_{x\in E} k(x)^*.$$

When k is a number field, the left hand arrow has torsion kernel, so it will suffice to construct elements in the kernel of the tame symbol. Also, R_q is trivial on symbols $\{f, c\}$, $c \in \mathbb{C}^*$, so it will suffice for our purposes to specify elements modulo the subgroup of $K_2\big(k(E)\big)$ generated by symbols with one element constant.

PROPOSITION 10.1.1. *Let f, g be functions on E defined over k. Let C be an integer and assume all points of order C are defined over k and that the divisors of f and g are supported in the group E_C of such points. Then there exist $f_i \in k(E)^*$, $c_i \in k^*$ such that*

$$\{f, g\}^C \cdot \prod \{f_i, c_i\} \in \operatorname{Ker}(\textit{tame symbol}).$$

PROOF. Consider the diagram

$$\begin{array}{ccccccc} k(E)^* \otimes k^* & \xrightarrow{\operatorname{div}\otimes \operatorname{Id}} & \coprod_{\substack{x\in E \\ \text{closed point}}} k^* & \longrightarrow & \operatorname{Pic}(E)\otimes k^* & \longrightarrow & 0 \\ \Big\downarrow{\scriptstyle \text{symbol}} & & \Big\downarrow & & & & \\ K_2\big(k(E)\big) & \xrightarrow{\text{tame}} & \coprod_{\substack{x\in E \\ \text{closed point}}} k(x)^*. & & & & \end{array} \tag{10.1.1}$$

Since f, g and all points of E_C are defined over k, we see

$$\operatorname{tame}\{f,g\} \in \coprod k^*.$$

Clearly $\operatorname{tame}\{f,g\}^C \to 0$ in $\operatorname{Pic}(E)\otimes k^*$, so modifying $\{f,g\}^C$ by something in the image of $k(E)^* \otimes k^*$ we get an element of Ker(tame). □

EXERCISE 10.1.2. Prove an analogous result without assuming all points of order C defined over k.

Let ρ denote a function on E with poles of order 1 at every non-zero point of E_C and a zero of order $C^2 - 1$ at the origin. Note that such a function exists when $E_C \cong (\mathbb{Z}/C)^{\oplus 2}$. For example, if $k \hookrightarrow \mathbb{C}$, $E(\mathbb{C}) = \mathbb{C}/[1,\tau]$, we have

$$\sum_{a,b=0}^{C-1} \frac{a+b\tau}{C} = \frac{C(C-1)}{2} + \frac{C(C-1)}{2}\tau \in [1,\tau].$$

For $a \in E_C$, $a \neq 0$, let f_a denote the function with zero of order C at a and pole of order C at 0. Define $S_a \in \operatorname{Ker}(\text{tame})$ by

$$S_a \equiv \{\rho, f_a\} \pmod{\operatorname{Image} k(E)^* \otimes k^*}. \tag{10.1.2}$$

S_a is actually well-defined modulo torsion and symbols with both entries constant, as one sees by looking at the kernel of $\operatorname{div}\otimes\operatorname{Id}$ in (10.1.1). When $K_2(k)$ is torsion (e.g., k a number field), S_a is well-defined modulo torsion.

10.2. Our main result can be given a purely analytic formulation as follows: fix $E = \mathbb{C}/[1,\tau]$ with $y = \operatorname{Im}\tau > 0$, $q = e^{2\pi i\tau}$, and let $C > 0$ be an integer. Let

$$f\colon \mathbb{Z}/C\mathbb{Z} \times \mathbb{Z}/C\mathbb{Z} \to \mathbb{C}$$

be an odd function, $f(-a,-b) = -f(a,b)$, and write

$$\hat{f}(k,\ell) = \frac{1}{C^2}\sum_{a,b=0}^{C-1} f(a,b)e^{2\pi i(-ak+b\ell)/C}. \tag{10.2.1}$$

THEOREM 10.2.1. *With notation as above,*

$$\sum_{k,\ell=0}^{C-1} \hat{f}(k,\ell) R_q(S_{\frac{k+\ell\tau}{C}}) = \frac{iy^2C^3}{\pi} \sum_{\substack{m,n\in\mathbb{Z}\\(m,n)\neq(0,0)}} \frac{f(m,n)}{(m\tau+n)^2(m\overline{\tau}+n)}.$$

The relationship between the right hand side and the zeta function of E in the complex multiplication case will be discussed in the next lecture. The remainder of this lecture is devoted to the proof of Theorem 10.2.1.

LEMMA 10.2.2. *We have*

$$R_q(S_{(k+\ell\tau)/C}) = C^3\left(R_q(e^{2\pi i(k+\ell\tau)/C}) + 4\pi^2y^2\left(\frac{\ell^3}{3C^3} - \frac{\ell^2}{2C^2} + \frac{\ell}{6C}\right)\right),$$

where R_q on the right denotes the function R_q and R_q on the left is the regulator.

PROOF. Since R_q is trivial on symbols with one element constant

$$\begin{aligned} R_q(S_{(k+\ell\tau)/C}) &= R_q\{\rho, f_{(k+\ell\tau)/C}\} \\ &= J_q\{\rho, f_{(k+\ell\tau)/C}\} + iD_q\big((\rho) * (f_{(k+\ell\tau)/C})\big). \end{aligned}$$

Writing $\sigma\colon \mathbb{C} \to E = \mathbb{C}/[1,\tau]$ and writing $\overline{D}_q$ for the function on E, we see

$$\begin{aligned} \overline{D}_q\big((\rho) * (f_{(k+\ell\tau)/C})\big) = C\sum_{\substack{Ca=0\\a\neq0}} &\left(-\overline{D}_q\left(-a+\sigma\left(\frac{k+\ell\tau}{C}\right)\right) + \overline{D}_q(-a)\right) \\ &+ C(C^2-1)\left(\overline{D}_q\left(\sigma\left(\frac{k+\ell\tau}{C}\right)\right) - \overline{D}_q(0)\right). \end{aligned}$$

Since $\overline{D}_q(0) = \sum_{n\in\mathbb{Z}} D(q^n) = 0$, the sum on the right collapses to

$$C\overline{D}_q\left(\sigma\left(\frac{k+\ell\tau}{C}\right)\right),$$

so

$$\overline{D}_q\big((\rho) * (f_{(k+\ell\tau)/C})\big) = C^3 D_q(e^{2\pi i(k+\ell\tau)/C}). \tag{10.2.2}$$

To evaluate $J_q(S_{(k+\ell\tau)/C})$ we choose liftings of (ρ) and $(f_(k+\ell\tau)/C)$ to divisors on $\mathbb{C}^*$ as follows

$$\begin{aligned} (\tilde{\rho}) &= -\sum_{a,b=0}^{C-1} (e^{2\pi i(a+b)/C}) + (e^{2\pi iC(C-1)\tau/2}) + (C^2-1)(1) \\ (\widetilde{f_{(k+\ell\tau)/C}}) &= C(e^{2\pi i(k+\ell\tau)/C}) - (C-1)(1) - (e^{2\pi i\ell\tau}). \end{aligned} \tag{10.2.3}$$

We write

$$A = J_q\Big[\Big(-\sum_{a,b=0}^{C-1}(e^{2\pi i(a+b\tau)/C}) + C^2(1)\Big) * (\widetilde{f_{(k+\ell\tau)/C}})\Big]$$

$$B = J_q\big[\big((e^{\pi i C(C-1)\tau}) - (1)\big)*(\widetilde{f_{(k+\ell\tau)/C}})\big]$$

so

$$J_q\big((\tilde{\rho}) * (\widetilde{f_{(k+\ell\tau)/C}})\big) = A + B.$$

Note

$$J_q(qx) - J_q(x) = -(\log|x|)^2$$

$$J_q(q^{C(C-1)/2}x) - J_q(x) = -\sum_{r=0}^{C(C-1)/2-1}(\log|xq^r|)^2$$

and if we replace x by a divisor $\mathcal{D} = \sum d_j(\alpha_j)$ on $\mathbb{C}^*$ with $\sum d_j = 0$, $\prod \alpha_j^{d_j} = 1$, we get

$$J_q(q^{C(C-1)/2} * \mathcal{D}) - J_q(\mathcal{D}) = -\frac{C(C-1)}{2}\sum_j d_j(\log|\alpha_j|)^2.$$

Take $\mathcal{D} = (\widetilde{f_{(k+\ell\tau)/C}})$, so

$$\sum d_j(\log|\alpha_j|)^2 = 4\pi^2 y^2\Big(\frac{\ell^2}{C} - \ell^2\Big) \tag{10.2.4}$$
$$B = 2\pi^2 y^2 \ell^2 (C-1)^2.$$

We now compute A.

$$\begin{aligned}
A &= C^3 J_q\big((e^{2\pi i((k+\ell\tau)/C)}) - (1)\big) \\
&\quad - C J_q\Big(\sum_{a,b=0}^{C-1}(e^{2\pi i[a+k+(b+\ell)\tau]/C}) - (e^{2\pi i(a+b)/C})\Big) \\
&\quad - C^2 J_q\big((e^{2\pi i\ell\tau}) - (1)\big) + J_q\Big(\sum_{a,b=0}^{C-1}(e^{2\pi i[(a+b\tau)/C+\ell\tau]}) - (e^{2\pi i(a+b\tau)/C})\Big) \\
&= A_1 - A_2 - A_3 + A_4.
\end{aligned}$$

Note $J_q(1) = 0$, so

$$A_1 = C^3 J_q(e^{2\pi i((k+\ell\tau)/C)})$$

$$\begin{aligned}
A_2 &= C J_q\Big(\sum_{a=0}^{C-1}\sum_{b=0}^{\ell-1}(e^{2\pi i[(a+b\tau)/C+\tau]}) - (e^{2\pi i(a+b\tau)/C})\Big) \\
&= -4\pi^2 y^2 C^2 \sum_{b=0}^{\ell-1}\frac{b^2}{C^2} = -2\pi^2 y^2 \frac{(\ell-1)\ell(2\ell-1)}{3}
\end{aligned}$$

$$\begin{aligned} A_3 &= C^2 J_q(e^{2\pi i \ell \tau}) \\ &= -4\pi^2 y^2 C^2 \sum_{b=0}^{\ell-1} b^2 = -2\pi^2 y^2 \frac{(\ell-1)\ell(2\ell-1)}{3} C^2 \\ A_4 &= -4\pi^2 y^2 C \sum_{b=0}^{C-1} \sum_{r=0}^{\ell-1} \left(r + \frac{b}{C} \right)^2 \\ &= \frac{-4\pi^2 y^2}{C} \sum_{s=0}^{C\ell-1} s^2 = \frac{-2\pi^2 y^2 (C\ell-1)\ell(2C\ell-1)}{3}. \end{aligned}$$

Combining these calculations,

$$\begin{aligned} &R_q(S_{(k+\ell\tau)/C}) \\ &= J_q\big((\tilde{\rho}) * (\widetilde{f_{(k+\ell\tau)/C}})\big) + i\overline{D}_q\big((\rho) * (f_{(k+\ell\tau)/C})\big) \\ &= A_1 - A_2 - A_3 + A_4 - B + iC^3 D_q(e^{2\pi i (k+\ell\tau)/C}) \\ &= C^3 R_q(e^{2\pi i (k+\ell\tau)/C}) - A_2 - A_3 + A_4 + B \\ &= C^3 R_q(e^{2\pi i (k+\ell\tau)/C}) + 2\pi^2 y^2 \bigg[\frac{(\ell-1)\ell(2\ell-1)}{3} + \frac{C^2(\ell-1)\ell(2\ell-1)}{3} \\ &\qquad\qquad - \frac{(C\ell-1)\ell(2C\ell-1)}{3} + \ell^2 (C-1)^2 \bigg] \\ &= C^3 R_q(e^{2\pi i (k+\ell\tau)/C}) + 4\pi^2 y^2 \left(\frac{\ell^3}{3C^3} - \frac{\ell^2}{2C^2} + \frac{\ell}{6C} \right). \qquad \square \end{aligned}$$

LEMMA 10.2.3. *Recall* $\hat{f}(k,\ell) = \big(\sum_{a,b=0}^{C-1} f(a,b) e^{2\pi i(-ak+b\ell)/C}\big)/C^2$. *We have*

$$4\pi^2 y^2 \sum_{k,\ell=0}^{C-1} \hat{f}(k,\ell) \left(\frac{\ell^3}{3c^3} - \frac{\ell^2}{2C^2} + \frac{\ell}{6C} \right) = \frac{iy^2}{\pi} \sum_{\substack{n\in\mathbb{Z} \\ n\neq 0}} \frac{f(0,n)}{n^3}.$$

PROOF. We use the well-known Fourier series

$$\sum_{\substack{n\in\mathbb{Z} \\ n\neq 0}} \frac{e^{2\pi i n x}}{n^3} = 4\pi^3 i \left(\frac{x^3}{3} - \frac{x^2}{2} + \frac{x}{6} \right)$$

to write

$$4\pi^2 y^2 \sum_{k,\ell} \hat{f}(k,\ell) \left(\frac{\ell^3}{3C^3} - \frac{\ell^2}{2C^2} + \frac{\ell}{6C} \right) = \frac{-y^2 i}{\pi} \sum_{k,\ell} \hat{f}(k,\ell) \sum_{\substack{n\in\mathbb{Z} \\ n\neq 0}} \frac{e^{2\pi i \ell n/C}}{n^3}$$

$$= \frac{-y^2 i}{\pi C^2} \sum_{a,b,k,\ell} f(a,b) e^{2\pi i(-ak+b\ell)/C} \sum_{n\neq 0} \frac{e^{2\pi i \ell n/C}}{n^3}$$

$$= \frac{-y^2 i}{\pi} \sum_{n\neq 0} \frac{f(0,-n)}{n^3} = \frac{y^2 i}{\pi} \sum \frac{f(0,n)}{n^3},$$

* because f is odd. □

10.3. Using Lemmas 10.2.2 and 10.2.3, the main result, Theorem 10.2.1 can be rewritten

$$\sum_{k,\ell=0}^{C-1} \hat{f}(k,\ell) R_q(e^{2\pi i(k+\ell\tau)/C}) = \frac{y^2 i}{\pi} \sum_{\substack{m,n\in\mathbb{Z}\\ m\neq 0}} \frac{f(m,n)}{(m\tau+n)^2(m\bar{\tau}+n)}. \tag{10.3.1}$$

For the proof, it will be convenient to write

$$\begin{aligned}
R_q(x) &= \sum_{n=0}^{\infty} \log|xq^n| \log|1-xq^n| - \sum_{n=1}^{\infty} \log|x^{-1}q^n| \log|1-x^{-1}q^n| \\
&\quad + i\sum_{n=0}^{\infty}\left(\log|xq^n| \arg(1-xq^n) - \operatorname{Im}\int_0^{xq^n} \log(1-t)\,\frac{dt}{t}\right) \\
&\quad - i\sum_{n=1}^{\infty}\left(\log|x^{-1}q^n| \arg(1-x^{-1}q^n) - \operatorname{Im}\int_0^{x^{-1}q^n} \log(1-t)\,\frac{dt}{t}\right) \\
&= \left(\sum_{n=0}^{\infty} \log|xq^n| \log(1-xq^n) - \sum_{n=1}^{\infty} \log|x^{-1}q^n| \log(1-x^{-1}q^n)\right) \\
&\quad - i\left(\sum_{n=0}^{\infty} \operatorname{Im}\int_0^{xq^n} \log(1-t)\,\frac{dt}{t} - \sum_{n=1}^{\infty} \operatorname{Im}\int_0^{x^{-1}q^n} \log(1-t)\,\frac{dt}{t}\right) \\
&= R'_q(x) - iR''_q(x).
\end{aligned}$$

(In what follows $|xq^n| < 1$ for $n \geq 0$ and $|x^{-1}q^n| < 1$ for $n \geq 1$ so there will be no ambiguity about branches of functions.) Put

$$\begin{aligned}
L &= \sum_{k,\ell=0}^{C-1} \hat{f}(k,\ell) R'_q(e^{2\pi i((k+\ell\tau)/C)}) \\
M &= -i\sum \hat{f}(k,\ell) R''_q(e^{2\pi i(k+\ell\tau)/C}).
\end{aligned} \tag{10.3.2}$$

PROPOSITION 10.3.1. *We have*

$$L = \frac{-y}{2\pi} \sum_{\substack{m\in\mathbb{Z}\\ m\neq 0}} \sum_{n\in\mathbb{Z}} \frac{f(m,n)}{m(m\tau+n)^2}.$$

PROOF.

$$L = -2\pi y \sum_{k,\ell=0}^{C-1} \hat{f}(k,\ell)\Bigg(\sum_{n=0}^{\infty}\left(n+\frac{\ell}{C}\right)\log\left(1-e^{2\pi i[(k+\ell\tau)/C+n\tau]}\right) - \sum_{n=1}^{\infty}\left(n-\frac{\ell}{C}\right)\log\left(1-e^{2\pi i[(-k-\ell\tau)/C+n\tau]}\right)\Bigg).$$

To consolidate the two series note that (replacing k, ℓ by $C-k$, $C-\ell$ and n by $n+1$),

$$\begin{aligned}
-\sum_{k,\ell=0}^{C-1} \hat{f}(k,\ell)\sum_{n=1}^{\infty}\left(n-\frac{\ell}{C}\right)&\log\left(1-e^{2\pi i[(-k-\ell\tau)/C+n\tau]}\right)\\
&= \sum_{k=0}^{C-1}\sum_{\ell=1}^{C}\hat{f}(k,\ell)\sum_{n=0}^{\infty}\left(n+\frac{\ell}{C}\right)\log\left(1-e^{2\pi i[(k+\ell\tau)/C+n\tau]}\right)\\
&= \sum_{k,\ell=0}^{C-1} \hat{f}(k,\ell)\sum_{n=0}^{\infty}\left(n+\frac{\ell}{C}\right)\log\left(1-e^{2\pi i[(k+\ell\tau)/C+n\tau]}\right)\\
&\qquad + \sum_{k=0}^{C-1}\hat{f}(k,C)\sum_{n=1}^{\infty} n\log\left(1-e^{2\pi i(k/C+n\tau)}\right)\\
&\qquad - \sum_{k=0}^{C-1}\hat{f}(k,0)\sum_{n=0}^{\infty} n\log\left(1-e^{2\pi i(k/C+n\tau)}\right)\\
&= \sum_{k,\ell=0}^{C-1} \hat{f}(k,\ell)\sum_{n=0}^{\infty}\left(n+\frac{\ell}{C}\right)\log\left(1-e^{2\pi i[(k+\ell\tau)/C+n\tau]}\right).
\end{aligned}$$

Thus

$$\begin{aligned}
L &= -4\pi y \sum_{k,\ell=0}^{C-1} \hat{f}(k,\ell)\sum_{n=0}^{\infty}\left(n+\frac{\ell}{C}\right)\log\left(1-e^{2\pi i[(k+\ell\tau)/C+n\tau]}\right)\\
&= \frac{4\pi y}{C^2}\sum_{k,\ell,a,b} f(a,b)e^{2\pi i[(-ak+b\ell)/C]}\sum_{n=0}^{\infty}\sum_{m=1}^{\infty}\left(n+\frac{\ell}{C}\right)\frac{e^{2\pi i m[(k+\ell\tau)/C-n\tau]}}{m}.
\end{aligned}$$

This can be rewritten

$$L = \frac{4\pi y}{C^2}\sum_{a,b} f(a,b)\sum_{\substack{n=0,m=1\\ m\equiv a \bmod C}}^{\infty}\frac{ne^{2\pi i[(m\tau+b)/C]}}{m}. \tag{10.3.3}$$

LEMMA 10.3.2. *For* $\operatorname{Im} x > 0$, $\sum_{n=0}^{\infty} ne^{2\pi i n x} = -(4\sin^2\pi x)^{-1}$.

PROOF.

$$\sum ne^{2\pi inx} = \frac{1}{2\pi i}\frac{d}{dx}\left(\frac{1}{1-e^{2\pi ix}}\right) = \frac{e^{2\pi ix}}{(1-e^{2\pi ix})^2}$$
$$= \frac{1}{(e^{\pi ix}-e^{-\pi ix})^2} = \frac{-1}{4\sin^2\pi x}.$$ □

Returning to the expression for L (10.3.3), we see

$$L = \frac{-\pi y}{C^2}\sum_{a,b} f(a,b) \sum_{\substack{m=1\\ m\equiv a \bmod C}}^{\infty} \frac{1}{m\sin^2\pi[(m\tau+b)C]}.$$

Now use

$$\frac{\pi^2}{\sin^2\pi x} = \sum_{n\in\mathbb{Z}} \frac{1}{(x+n)^2}$$

to write

$$L = \frac{-y}{\pi C^2}\sum_{a,b} f(a,b) \sum_{\substack{m=1\\ m\equiv a \bmod C}}^{\infty} \sum_{n\in\mathbb{Z}} \frac{1}{m[(m\tau+b)/C+n]^2}$$
$$= \frac{-y}{2\pi}\sum_{\substack{m,n=-\infty\\ m\neq 0}}^{\infty} \frac{f(m,n)}{m(m\tau+n)^2}.$$

This proves Proposition 10.3.1. □

PROPOSITION 10.3.3. *Let M be as in (10.3.2) Then*

$$M = \frac{-1}{2\pi}\sum_{\substack{m,n=-\infty\\ m\neq 0}}^{\infty} f(m,n)\,\mathrm{Im}\left(\frac{1}{m^2(m\tau+n)}\right).$$

PROOF. For $|x| < 1$, the "natural" branch of $\int_0^x \log(1-t)\,dt/t$ is $-\sum_{m=1}^{\infty} x^m/m^2$. Thus

$$M = i\sum_{k,\ell=0}^{C-1} \hat{f}(k,\ell)\,\mathrm{Im}\left(\sum_{\substack{n=0\\ m=1}}^{\infty} \frac{e^{2\pi i[(k+\ell\tau)/C+n\tau]m}}{m^2} - \sum_{\substack{n=1\\ m=1}}^{\infty} \frac{e^{2\pi i[(-k-\ell\tau)/C+n\tau]m}}{m^2}\right).$$

As with L, it will be convenient to combine the two sums:

$$-\sum_{k,\ell=0}^{C-1}\hat{f}(k,\ell)\sum_{m=n=1}^{\infty}\frac{e^{2\pi i m[(k+\ell\tau)/C+n\tau]}}{m^2}$$
$$=\sum_{k=0}^{C-1}\sum_{\ell=1}^{C}\hat{f}(k,\ell)\sum_{\substack{m=1\\n=0}}^{\infty}\frac{e^{2\pi i m[(k+\ell\tau)/C+n\tau]}}{m^2}$$
$$=\sum_{k,\ell=0}^{C-1}\hat{f}(k,\ell)\sum_{\substack{m=1\\n=0}}^{\infty}\frac{e^{2\pi i m[(k+\ell\tau)/C+n\tau]}}{m^2}-\sum_{k=0}^{C-1}\hat{f}(k,0)\sum_{m=1}^{\infty}\frac{e^{2\pi i mk/C}}{m^2},$$

so

$$M=2i\sum_{k,\ell=0}^{C-1}\hat{f}(k,\ell)\operatorname{Im}\left(\sum_{\substack{n=0\\m=1}}^{\infty}\frac{e^{2\pi i[(k+\ell\tau)/C+n\tau]m}}{m^2}\right) -i\sum_{k=0}^{C-1}\hat{f}(k,0)\operatorname{Im}\left(\sum_{m=1}^{\infty}\frac{e^{2\pi i mk/C}}{m^2}\right)$$
$$=M_1+M_2.$$

LEMMA 10.3.4.

$$M_1=\frac{1}{C}\sum_{b=0}^{C-1}\sum_{m=1}^{\infty}\frac{f(m,b)}{m^2}-\frac{1}{2\pi}\sum_{\substack{m,n=-\infty\\m\neq 0}}^{\infty}f(m,n)\operatorname{Im}\left(\frac{1}{m^2(m\tau+n)}\right).$$

PROOF.

$$M_1=\frac{2i}{C^2}\sum_{a,b,k,\ell}f(a,b)e^{2\pi i[(-ak+b\ell)/C]}\operatorname{Im}\left(\sum_{\substack{=0\\m=1}}^{\infty}\frac{e^{2\pi i[(k+\ell\tau)/C+n\tau)]m}}{m^2}\right).$$

Note $\sum_{a,b}f(a,b)\operatorname{Re}(e^{2\pi i[(-ak+b\ell)/C]})=0$ because $f(a,b)=-f(-a,-b)$, and also

$$\operatorname{Im}\left(ie^{2\pi i[(-ak+b\ell)/C]}(*)\right)$$
$$=-\operatorname{Im}\left(e^{2\pi i[(-ak+b\ell)/C]}\right)\operatorname{Im}(*)+\operatorname{Re}\left(e^{2\pi i[(-ak+b\ell)/C]}\right)\operatorname{Re}(*).$$

It follows that

$$M_1=\frac{2}{C^2}\sum_{a,b,k,\ell}f(a,b)\operatorname{Im}\left(i\sum_{\substack{n=0\\m=1}}^{\infty}\frac{e^{2\pi i[(m-a)k/C+(nC+\ell)(m\tau+b)/C]}}{m^2}\right)$$
$$=\frac{2}{C}\sum_{a,b}f(a,b)\operatorname{Im}\left(i\sum_{\substack{m=1\\m\equiv a \bmod C}}^{\infty}\sum_{n=0}^{\infty}\frac{e^{2\pi i n(m\tau+b)/C}}{m^2}\right).$$

Using the identity

$$\frac{1}{1-e^{2\pi i x}} = \frac{i\cot\pi x + 1}{2}$$

we can rewrite

$$M_1 = \frac{1}{C}\sum_{a,b} f(a,b)\,\mathrm{Im}\Bigg(\sum_{\substack{m=1\\ m\equiv a \bmod C}}^{\infty} \frac{-\cot\pi([m\tau+b]/C)+i}{m^2}\Bigg)$$

$$= \frac{1}{C}\sum_{\substack{m=1\\ b}}^{\infty}\frac{f(m,b)}{m^2} - \frac{1}{C}\sum_{a,b} f(a,b)\,\mathrm{Im}\Bigg(\sum_{\substack{=1\\ m\equiv a}}^{\infty}\frac{1}{m^2}\cot\Big(\pi\Big(\frac{m\tau+b}{C}\Big)\Big)\Bigg).$$

Note

$$\cot\pi x = \frac{1}{\pi}\lim_{N\to\infty}\sum_{n=-N}^{N}\frac{1}{x+n},$$

and

$$\mathrm{Im}(\cot\pi x) = \frac{1}{\pi}\sum_{n\in\mathbb{Z}}\mathrm{Im}\Big(\frac{1}{x+n}\Big)$$

(the series converging without any special sort of summation), so we get

$$M_1 = \frac{1}{C}\sum_{\substack{m=1\\ b}}^{\infty}\frac{f(m,b)}{m^2} - \frac{1}{\pi C}\sum_{a,b} f(a,b)\sum_{\substack{m=1\\ m\equiv a}}^{\infty}\sum_{n\in\mathbb{Z}}\mathrm{Im}\Big(\frac{1}{m^2([m\tau+b]/C+n)}\Big)$$

$$= \frac{1}{C}\sum_{\substack{m=1\\ b}}^{\infty}\frac{f(m,b)}{m^2} - \frac{1}{2\pi}\sum_{\substack{m\in\mathbb{Z}\\ m\neq 0}}\sum_{n\in\mathbb{Z}} f(m,n)\,\mathrm{Im}\Big(\frac{1}{m^2(m\tau+n)}\Big).$$

This completes the computation of M_1. □

LEMMA 10.3.5.

$$M_2 = \frac{-1}{C}\sum_{m=1}^{\infty}\sum_{b=0}^{C-1}\frac{f(m,b)}{m^2}.$$

PROOF.

$$M_2 = -i\sum_{k=0}^{C-1}\hat{f}(k,0)\,\mathrm{Im}\Bigg(\sum_{m=1}^{\infty}\frac{e^{2\pi i\frac{mk}{C}}}{m^2}\Bigg)$$

$$= \frac{-i}{C^2}\sum_{k,a,b} f(a,b)e^{-2\pi i ak/C}\,\mathrm{Im}\Bigg(\sum_{m=1}^{\infty}\frac{e^{2\pi i\frac{mk}{C}}}{m^2}\Bigg)$$

$$= \frac{-1}{C^2}\sum_{k,a,b} f(a,b)\,\mathrm{Im}\Bigg(i\sum_{m=1}^{\infty}\frac{e^{2\pi i\frac{k}{C}(m-a)}}{m^2}\Bigg) = \frac{-1}{C}\sum\frac{f(m,b)}{m^2}.$$ □

We return now to the proof of Proposition 10.3.3.

$$M = M_1 + M_2 = \frac{-1}{2\pi} \sum_{m\neq 0} \sum_{n\in\mathbb{Z}} f(m,n) \operatorname{Im}\left(\frac{1}{m^2(m\tau + n)} \right)$$

as desired. □

We can now complete the proof of Theorem 10.2.1:

$$\sum_{k,\ell=0}^{C-1} \hat{f}(k,\ell) R_q(e^{2\pi i(k+\ell\tau)/C}) = L + M$$

$$= \frac{-y}{2\pi} \left(\sum_{\substack{m\in\mathbb{Z} \\ m\neq 0}} \sum_{n\in\mathbb{Z}} f(m,n) \left(\frac{1}{m(m\tau+n)^2} + \frac{1}{y} \operatorname{Im} \frac{1}{m^2(m\tau+n)} \right) \right).$$

But

$$\frac{1}{m(m\tau+n)^2} + \frac{1}{y}\operatorname{Im}\frac{1}{m^2(m\tau+n)}$$

$$= \frac{1}{m(m\tau+n)^2} + \frac{1}{2iy}\left(\frac{1}{m^2(m\tau+n)} - \frac{1}{m^2(m\overline{\tau}+n)} \right)$$

$$= \frac{1}{m(m+\tau n)^2} - \frac{1}{m(m\tau+n)(m\overline{\tau}+n)}$$

$$= -2iy \frac{1}{(m\tau+n)^2(m\overline{\tau}+n)}$$

so

$$\sum_{k,\ell=0}^{C-1} \hat{f}(k,\ell) R_q(e^{2\pi i(k+\ell\tau)/C}) = \frac{y^2 i}{\pi} \sum_{\substack{m,n\in\mathbb{Z} \\ m\neq 0}} \frac{f(m,n)}{(m\tau+n)^2(m\overline{\tau}+n)}.$$

This completes the proof of (10.3.1) and also Theorem 10.2.1. □

LECTURE 11

A Regulator Formula

11.1. In this final lecture, we will show how Theorem 10.3.2 leads to a regulator formula for the value at $s = 2$ of the zeta function of an elliptic curve E with complex multiplication by the ring of integers in an imaginary quadratic field κ with class number 1. We begin with some general lemmas, valid without hypotheses on E. Notation and hypotheses will be as in Lecture 10.

LEMMA 11.1.1.

$$R_q(S_{(a+b\tau)/C}) = -R_q(S_{(-a-b\tau)/C}).$$

PROOF. The divisor (ρ) (10.1.2) is invariant under the automorphism of E given by multiplication by -1, so $2(\rho)$ can be lifted to a $\mathbb{C}^*$-divisor $\sum d_j(\alpha_j)$ invariant under $\alpha \to \alpha^{-1}$ and such that $\sum d_j = 0$, $\prod \alpha_j^{d_j} = 1$. Let $\sum e_k(\beta_k)$ be a lifting of $C\big(([a + b\tau]/C) - (0)\big)$ with $\sum e_k = 0$, $\prod \beta_k^{e_k} = 1$. Then $\sum e_k(\beta_k^{-1})$ lifts $C\big(([-a - b\tau]/C) - (0)\big)$. We must show

$$\sum d_je_k\big(J_q(\alpha_j^{-1}\beta_k) + J_q(\alpha_j\beta_k^{-1})\big) = 0$$
$$= \sum d_je_k\big(D_q(\alpha_j^{-1}\beta_k) + D_q(\alpha_j\beta_k^{-1})\big).$$

It follows from Lemma 6.2.5 that $D_q(x^{-1}) = -D_q(x)$, so the right hand equality is clear. For the left hand, we know $J_q(x)+J_q(x^{-1}) = (\log|x|)^2$ from which the desired result follows. □

It is convenient to alter slightly the definition of the Fourier transform employed in Lecture 10. We write $f(a + b\tau)$ in place of $f(a, b)$, and define

$$\hat{f}(k + \ell\tau) = \frac{1}{C}\sum_{a,b=0}^{C-1} f(a + b\tau)e^{2\pi i[(-a\ell+bk)/C]}. \tag{11.1.1}$$

(Note the interchange of k and ℓ.) Our fundamental result, Theorem 10.2.1, becomes

$$\sum_{k,\ell=0}^{C-1} \hat{f}(k+\ell\tau) R_q(S_{(k+\ell\tau)/C}) = \frac{iy^2C^4}{\pi} \sum_{(m,n)\neq(0,0)} \frac{f(m+n\tau)}{(m+n\tau)^2(m+n\overline{\tau})}. \tag{11.1.2}$$

LEMMA 11.1.2.

$$R_q(S_{(a+b\tau)/C}) = \frac{y^2C^3}{\pi} \sum_{(m,n)\neq(0,0)} \frac{\sin(2\pi[an-bm]/C)}{(m+n\tau)^2(m+n\overline{\tau})}$$

PROOF. Let $g_{a+b\tau}(k+\ell\tau) = e^{2\pi i(a\ell-bk)/C}/C$. Then

$$\begin{aligned} \hat{g}_{a+b\tau}(m+n\tau) &= \frac{1}{C^2}\sum_{k,\ell} e^{2\pi i(a\ell-bk)/C} e^{2\pi i(-kn+\ell m)/C} \\ &= \frac{1}{C^2}\sum_{k,\ell} e^{2\pi i[\ell(a+m)-k(b+n)]/C} \\ &= \begin{cases} 0 & a\neq -m \text{ or } b\neq -n \\ 1 & a=-m, b=-n \end{cases} \\ &= \delta_{-a-b\tau}(m+n\tau) \end{aligned}$$

with obvious notation. Writing $f_{a+b\tau} = g_{-a-b\tau} - g_{a+b\tau}$ we get

$$\begin{aligned} R_q(S_{(a+b\tau)/C}) &= \frac{1}{2}\sum \hat{f}_{a+b\tau}(k+\ell\tau) R_q(S_{(k+\ell\tau)/C}) \\ &= \frac{iy^2C^4}{2\pi}\sum \frac{f_{a+b\tau}(m+n\tau)}{(m+n\tau)^2(m+n\overline{\tau})} \\ &= \frac{y^2C^3}{\pi}\sum \frac{\sin(2\pi[an-bm]/C)}{(m+n\tau)^2(m+n\overline{\tau})}. \end{aligned}$$

□

The following description of the "modular behavior" of R_q seemed worth including although it will not be used:

LEMMA 11.1.3. *Let $\tau' = (\alpha\tau+\beta)/(\gamma\tau+\delta)$ with $\left(\begin{smallmatrix}\alpha & \gamma \\ \beta & \delta\end{smallmatrix}\right) \in \mathrm{SL}_2(\mathbb{Z})$. Write $q' = e^{2\pi i\tau'}$ and let $\left(\begin{smallmatrix} b \\ a\end{smallmatrix}\right) = \left(\begin{smallmatrix}\alpha & \gamma \\ \beta & \delta\end{smallmatrix}\right)\left(\begin{smallmatrix} b' \\ a'\end{smallmatrix}\right)$. Then*

$$R_{q'}(S_{(a'+b'\tau')/C}) = (\gamma\tau+\delta)^2(\gamma\overline{\tau}+\delta) R_q(S_{(a+b\tau)/C}).$$

PROOF. Since $\mathrm{SL}_2(\mathbb{Z})$ preserves the symplectic form $\left(\begin{smallmatrix} 0 & 1 \\ -1 & 0\end{smallmatrix}\right)$, we find $-a'\ell' + b'k' = -a\ell + bk$, where $\left(\begin{smallmatrix} \ell \\ k\end{smallmatrix}\right) = \left(\begin{smallmatrix}\alpha & \gamma \\ \beta & \delta\end{smallmatrix}\right)\left(\begin{smallmatrix} \ell' \\ k'\end{smallmatrix}\right)$. Thus

$$R_{q'}(S_{(a'+b'\tau')/C}) = \frac{yC^3}{\pi}\sum_{k',\ell'}\frac{\sin(2\pi[a'\ell'-b'k']/C)}{(k'+\ell'\tau')^2(k'+\ell'\overline{\tau}')}$$
$$= \frac{yC^3}{\pi}\sum_{k',\ell'}\frac{\sin(2\pi[a\ell-bk]/C)}{(k'+\ell'\tau')^2(k'+\ell'\overline{\tau}')}.$$

Note

$$k'+\ell'\tau' = (\gamma\tau+\delta)^{-1}(k'\delta+\ell'\beta+(k'\gamma+\ell'\alpha)\tau) = (\gamma\tau+\delta)^{-1}(k+\ell\tau)$$

whence

$$R_{q'}(S_{(a'+b'\tau')/C}) = \frac{yC^3(\gamma\tau+\delta)^2(\gamma\overline{\tau}+\delta)}{\pi}\sum_{k,\ell}\frac{\sin(2\pi[a\ell-bk]/C)}{(k+\ell\tau)^2(k+\ell\overline{\tau})}$$
$$= (\gamma\tau+\delta)^2(\gamma\overline{\tau}+\delta)R_q(S_{(a+b\tau)/C}). \qquad \square$$

It is convenient to denote by $\langle\ ,\ \rangle$ the bilinear form

$$\langle\ ,\ \rangle\colon \mathcal{O}/C\mathcal{O}\times\mathcal{O}/C\mathcal{O}\to\mathbb{C}^*$$
$$\langle a+b\tau, k+\ell\tau\rangle = e^{2\pi i(-a\ell+bk)/C}.$$

We assume henceforth that $\mathcal{O} = \mathbb{Z}+\mathbb{Z}\tau$ is an order in an imaginary quadratic field κ.

LEMMA 11.1.4. *$\langle xy, z\rangle = \langle x, \bar{y}z\rangle$, where $\bar{y}$ is the complex conjugate.*

PROOF. Write $x = x_1+x_2\tau$, $y = y_1+y_2\tau$, $z = z_1-z_2\tau$, and suppose $\tau^2+A\tau+B=0$, with x_i, y_i, z_i, A, $B\in\mathbb{Z}$. Then

$$xy = x_1y_1 - x_2y_2B + (x_1y_2+x_2y_1-x_2y_2A)\tau$$
$$\langle xy, z\rangle = \exp\left(2\pi i\left(\frac{x_2y_2Bz_2-x_1y_1z_2+x_1y_2z_1+x_2y_1z_1-x_2y_2Az_1}{C}\right)\right)$$
$$\bar{y}z = y_1z_1+y_2z_2B+y_2z_1\overline{\tau}+y_1z_2\tau$$
$$= (y_1z_1+y_2z_2B-y_2z_1A)+(y_1z_2-y_2z_1)\tau$$
$$\langle x, yz\rangle = \exp\left(2\pi i\left(\frac{-x_1y_1z_2+x_1y_2z_1+x_2y_1z_1+x_2y_2z_2B-x_2y_2z_1A}{C}\right)\right)$$

$\square$

COROLLARY 11.1.5. *If $\zeta\in\mathcal{O}_\kappa$ is a unit (i.e., a root of 1) then*

$$R_q(S_{\zeta(a+b\tau)/C}) = \zeta^{-1}R_q(S_{(a+b\tau)/C}).$$

PROOF.

$$\begin{aligned}
R_q(S_{\zeta(a+b\tau)/C}) &= \frac{y^2C^3}{\pi}\sum\frac{\operatorname{Im}\langle\zeta(a+b\tau),k+\ell\tau\rangle}{(k+\ell\tau)^2(k+\ell\overline{\tau})}\\
&= \frac{y^2C^3}{\pi}\sum\frac{\operatorname{Im}\langle a+b\tau,\overline{\zeta}(k+\ell\tau)\rangle}{(k+\ell\tau)^2(k+\ell\overline{\tau})}\\
&= \frac{y^2C^3}{\pi}\sum\frac{\operatorname{Im}\langle a+b\tau,k'+\ell'\tau\rangle}{\zeta(k'+\ell'\tau)^2(k'+\ell'\overline{\tau})}\\
&= \zeta^{-1}R_q(S_{(a+b\tau)/C}).
\end{aligned}$$

□

Assume now for simplicity that $\mathcal{O}$ is the ring of integers in κ, and κ has class number 1.

COROLLARY 11.1.6. *Suppose $C = fg$, f, $g \in \mathcal{O}_\kappa$. Then*

$$\bar{g}\sum_{\mu\in\mathcal{O}/g\mathcal{O}} R_q(S_{(a+b\tau)/C+f\mu/C}) = R_q(S_{(a+b\tau)/f}).$$

PROOF. We have

$$\sum_{\mu\in\mathcal{O}/g\mathcal{O}}\langle\mu,\bar{f}(k+\ell\tau)\rangle = \begin{cases} N(g) & \bar{g}\mid k+\ell\tau\\ 0 & \text{otherwise.}\end{cases}$$

Thus

$$\begin{aligned}
\sum_{\mu\in\mathcal{O}/g\mathcal{O}} & R_q(S_{(a+b\tau)/C+f\mu/C})\\
&= \sum_{k,\ell}\frac{y^2C^3}{\pi}\sum_{\mu}\langle\mu,\bar{f}(k+\ell\tau)\rangle\frac{\operatorname{Im}\langle a+b\tau,k+\ell\tau\rangle}{(k+\ell\tau)^2(k+\ell\overline{\tau})}\\
&= N(g)\frac{y^2C^3}{\pi}\sum_{r,s}\frac{\operatorname{Im}\langle a+b\tau,\bar{g}(r+s\tau)\rangle}{(r+s\tau)^2(r+s\overline{\tau})\bar{g}^2g}\\
&= \frac{y^2C^3}{\bar{g}\pi}\sum_{r,s}\frac{\operatorname{Im}\langle g(a+b\tau),r+s\tau\rangle}{(r+s\tau)^2(r+s\overline{\tau})}\\
&= (\bar{g})^{-1}R_q(S_{g(a+b\tau)/C})\\
&= (\bar{g})^{-1}R_q(S_{(a+b\tau)/f}).
\end{aligned}$$

□

LEMMA 11.1.7. *Let χ have conductor $f \mid C$, $C = fg$. Then*

(i) *$\widehat{\chi}(x) = 0$ unless $C \mid \bar{f}x$.*
(ii) *) For $x \in (\mathcal{O}/C\mathcal{O})^*$, $\widehat{\chi}(xy) = \overline{\chi}(\bar{x})\widehat{\chi}(y)$.*
(iii) *Let $f_1 \parallel f$ and suppose $C \mid \bar{f}_1x$. Then $\widehat{\chi}(x) = 0$.*

PROOF. (i) We have

$$\begin{aligned}\widehat{\chi}(x) &= \frac{1}{C}\sum_{y\in\mathcal{O}/C\mathcal{O}}\chi(y)\langle x,y\rangle = \frac{1}{C}\sum_{y' \bmod f}\chi(y')\sum_{y'' \bmod g}\langle x,y'\rangle\langle x,fy''\rangle\\ &= \frac{1}{C}\sum_{y' \bmod f}\chi(y')\langle x,y'\rangle\sum_{y'' \bmod g}\langle x\bar{f},y''\rangle\\ &= \begin{cases}0 & \bar{g}\nmid x\\ N(g)(\sum_{y' \bmod f}\chi(y')\langle x,y'\rangle)/C & \bar{g}\mid x,\end{cases}\end{aligned}$$

where y' (resp. y'') run through representatives in $\mathcal{O}$ for the congruence classes modulo f (resp. g).

(ii) If $x\in(\mathcal{O}/C\mathcal{O})^*$,

$$\widehat{\chi}(xy) = \frac{1}{C}\sum_{z\in\mathcal{O}/C\mathcal{O}}\chi(z)\langle xy,z\rangle = \frac{1}{C}\overline{\chi}(\bar{y})\sum_{z}\chi(z\bar{y})\langle x,\bar{y}z\rangle = \overline{\chi}(\bar{y})\widehat{\chi}(x).$$

(iii) Since $f_1 \parallel f =$ conductor χ, we can find $y\equiv 1 \bmod \bar{f}_1$, y a unit in $\mathcal{O}/C\mathcal{O}$, with $\chi(\bar{y})\neq 1$. Since $C\mid \bar{f}_1\cdot x$

$$\widehat{\chi}(x) = \widehat{\chi}(xy) = \overline{\chi}(\bar{y})\widehat{\chi}(x).$$

Thus $\widehat{\chi}(x) = 0$. □

11.2. We now have the tools we need for the regulator formula. Assume as above that $\mathcal{O}=\mathbb{Z}+\mathbb{Z}\tau$ is

the ring of integers in an imaginary quadratic number field κ of class number 1. We fix an embedding $\kappa\hookrightarrow\mathbb{C}$, an integer C, and a character χ of $(\mathcal{O}/C\mathcal{O})^*$ which restricts to the given embedding $\mu_\kappa\hookrightarrow\mathbb{C}^*$ on the roots of 1. Let χ^{Gross} denote the Grössencharakter

$$\chi^{\mathrm{Gross}}(\mathfrak{p}) = \bar{h}\chi(h),\quad \mathfrak{p}=(h),\mathfrak{p}\nmid C.$$

Let f generate the conductor ideal of χ and write $C=fg$. From (11.1.2), Corollary 11.1.5, Corollary 11.1.6, and Lemma 11.1.7 we get

$$\begin{aligned}L(2,\chi^{\mathrm{Gross}}) &= \frac{\pi}{iy^2C^4}\sum_{w\in\mathcal{O}/C\mathcal{O}}\widehat{\chi}(w)R_q(S_{w/C})\\ &= \frac{\pi}{iy^2C^4}\sum_{x\in(\mathcal{O}/\bar{f}\mathcal{O})^*}\widehat{\chi}(x\bar{g})R_q(S_{x\bar{f}^{-1}})\\ &= \frac{\pi\widehat{\chi}(\bar{g})}{iy^2C^4}\sum_{x\in(\mathcal{O}/\bar{f}\mathcal{O})^*}\overline{\chi}(\bar{x})R_q(S_{x\bar{f}^{-1}})\end{aligned}\tag{11.2.1}$$

$$= \frac{\pi\widehat{\chi}(\bar{g})g}{iy^2C^4} \sum_{x\in(\mathcal{O}/C\mathcal{O})^*} \overline{\chi}(\bar{x})R_q(S_{x/C})$$
$$= \frac{\pi|\mu_\kappa|\widehat{\chi}(\bar{g})g}{iy^2C^4} \sum_{x\in(\mathcal{O}/C\mathcal{O})^*/\mu_\kappa} \overline{\chi}(\bar{x})R_q(S_{x/C}).$$

The j-invariant $j(E)$ is known to be real and to generate the Hilbert class field of κ. Since κ has class number 1, $j(E) \in \mathbb{Q}$ so we can choose a model $E_{\mathbb{Q}}$ defined over $\mathbb{Q}$. Deuring's theory associates to $E_{\mathbb{Q}}$ a Grössencharakter χ^{Gross} of κ with values in κ^*,

$$\chi^{\text{Gross}}(\mathfrak{p}) = \bar{x}\chi(x),$$

where $(x) = \mathfrak{p} \nmid C$. The character χ takes values in μ_κ and $\chi(\bar{x}) = \overline{\chi(x)}$ (this follows from [**Lan73**, Theorem 10, p. 140], which implies $\overline{\chi^{\text{Gross}}(\mathfrak{p})} = \chi^{\text{Gross}}(\overline{\mathfrak{p}})$,) so we can rewrite (11.2.1)

$$L(2,\chi^{\text{Gross}}) = \frac{\pi|\mu_\kappa|\widehat{\chi}(\bar{g})g}{iy^2C^4} R_q\left(\sum_{(\mathcal{O}/C\mathcal{O})^*/\mu_\kappa} S_{\frac{x\overline{\chi}(x)}{C}}\right) \tag{11.2.2}$$

The ray class group $(\mathcal{O}/C\mathcal{O})^*/\mu_\kappa$ acts on points of order C by

$$x \cdot \left(\frac{y}{C}\right) = \frac{x^{-1}\chi(x)y}{C}$$

and conjugation acts on these points in the natural way. The element

$$U \stackrel{\text{dfn}}{=} \sum_{(\mathcal{O}/C\mathcal{O})^*/\mu_\kappa} S_{x\overline{\chi}(x)/C} \tag{11.2.3}$$

is invariant under both these actions, and hence lies in

$$K_2(E_{\mathbb{Q}}) \otimes \mathbb{Q}.$$

(This follows from the existence of a norm map $K_2(E_L) \to K_2(E_{\mathbb{Q}})$ for $L/\mathbb{Q}$ finite.) We get

THEOREM 11.2.1. *With notations as above, let χ^{Gross} be the Grössencharakter associated to $E_{\mathbb{Q}}$. Let f be a generator for the conductor ideal and let $fg = C \in \mathbb{Z}$, $g \in \mathcal{O}$. Then there exists $U \in K_2(E_{\mathbb{Q}}) \otimes \mathbb{Q}$ (defined as above) such that*

$$L(2,\chi^{\text{Gross}}) = \frac{\pi|\mu_\kappa|\widehat{\chi}(\bar{g})g}{iy^2C^4} R_q(U). \tag{11.2.4}$$

REMARK 11.2.2. (i) $L(s,\chi^{\text{Gross}})$ is related to the zeta function of $E_{\mathbb{Q}}$, $\zeta_{E_{\mathbb{Q}}}$ (defined to be the product over all $p \in \mathbb{Z}$

such that E has non-degenerate reduction mod$p$ of the zeta function of the corresponding curve over $\mathbb{F}_p$), by

$$\zeta_{E_{\mathbb{Q}}}(s) = \zeta_{\mathbb{Q}}(s)\zeta_{\mathbb{Q}}(s-1)L(s,\chi^{\mathrm{Gross}})^{-1}.$$

(ii) It is presumably possible to renormalize the additive character defining $\widehat{\chi}$ so as to remove C and g from (11.2.4) and obtain a formula solely in terms of the conductor of χ. If the conductor ideal comes from $\mathbb{Z}$, we can take $g = 1$ and get such a formula immediately.

Since $L(s,\chi^{\mathrm{Gross}})$ has a product expansion converging for $\operatorname{Re} s > \frac{3}{2}$, we see from (11.2.4)

COROLLARY 11.2.3. *The element $U \in K_2(E_{\mathbb{Q}}) \otimes \mathbb{Q}$ defined above is non-zero.*

In view of Borel's work, the following conjecture seems irresistible:

CONJECTURE 11.2.4. *U spans $K_2(E_{\mathbb{Q}}) \otimes \mathbb{Q}$.*

More generally, for any elliptic curve over a number field one could conjecture that $\operatorname{rk} K_2(E) =$ order of zero of $L(E,s)$ at $s = 0$, assuming of course the existence of a functional equation relating $L(E,s)$ and $L(E,2-s)$.

Bibliography

[Bas71] H. Bass, *K_2 des corps globaux [d'après J. Tate, H. Garland,...]*, Exp. No. 394, Séminaire Bourbaki, 23e année (1970/1971), Lecture Notes in Math., vol. 244, 1971, pp. 233–255.

[Blo78] S. Bloch, *Applications of the dilogarithm function in algebraic K-theory and algebraic geometry*, Proceedings of the International Symposium on Algebraic Geometry (Kyoto, 1977) (M. Nagata, ed.), Kinokuniya , Tokyo, 1978, pp. 103–114.

[Bor53] A. Borel, *Sur la cohomologie des espaces fibrés principaux et des espaces homogènes de groupes de Lie compacts*, Ann. of Math. (2) **57** (1953), 115–207.

[Bor74] A. Borel, *Stable real cohomology of arithmetic groups*, Ann. Sci. École Norm. Sup. (4) **7** (1974), 235–272.

[Bor77] A. Borel, *Cohomologie de SL_n et valeurs de fonctions zeta aux points entiers*, Ann. Scuola Norm. Sup. Pisa Cl. Sci. (4) **4** (1977), no. 4, 613–636.

[BS66] Z. I. Borevich and I. R. Shafarevich, *Number theory*, Pure Appl. Math., vol. 20, Academic Press, New York-London, 1966.

[BSD65] B. Birch and H. P. F. Swinnerton-Dyer, *Notes on elliptic curves*. II, J. Reine Angew. Math. **218** (1965), 79–108.

[Cox35] H. S. M. Coxeter, *The functions of Schläfli and Lobatschevsky*, Quart. J. Math. Oxford Ser. **6** (1935), 13–29.

[Hel62] S. Helgason, *Differential geometry and symmetric spaces*, Pure Appl. Math., vol. 12, Academic Press, New York-London, 1962.

[HM62] G. Hochschild and G. D. Mostow, *Cohomology of Lie groups*, Illinois J. Math **6** (1962), 367–401.

[Keu78] F. Keune, *The relativization of K_2*, J. Algebra **54** (1978), no. 1, 159–177.

[Lan73] S. Lang, *Elliptic functions*, Addison-Wesley, Reading, MA-London-Amsterdam, 1973.

[Ler51] J. Leray, *Sur l'homologie des groupes de Lie, des espaces homogènes, et des espaces fibrés principaux*, Colloque de topologie (espace fibrés) (Bruxelles, 1950), G. Thone, Liège, 1951, pp. 101–115.

[Lic73] S. Lichtenbaum, *Values of zeta-functions, étale cohomology, and algebraic K-theory*, Algebraic K-Theory. II. "Classical" Algebraic K-Theory and Connections with Arithmetic (Seattle, WA, 1972) (H. Bass, ed.), Lecture Notes in Math., vol. 342, Springer-Verlag, Berlin-New York, 1973, pp. 489–501.

[Moo76] C. C. Moore, *Group extensions and cohomology*. III, Trans. Amer. Math. Soc. **221** (1976), no. 1, 1–33.

[Qui73] D. Quillen, *Higher algebraic K-theory*. I, Algebraic K-Theory. I. Higher K-Theories (Seattle, WA, 1972) (H. Bass, ed.), Lecture Notes in Math., vol. 341, Springer-Verlag, Berlin-New York, 1973, pp. 85–147.

[Rog07] L. J. Rogers, *On function sums connected with the series* $\sum x^n/n^2$, Proc. London Math. Soc. (2) **4** (1907), 169–189.

[Tat71] J. Tate, *Symbols in arithmetic*, Actes du Congrès International des Mathématiciens (Nice, 1970), vol. 1, Gauthier-Villars, Paris, 1971, pp. 201–211.

[vE53] W. I. van Est, *Group cohomology and Lie algebra cohomology in Lie groups*. I, Nederl. Akad. Wetensch. Proc. Ser. A **56** (1953), 484–492, II, 493–504.

[Wei61] A. Weil, *Adeles and algebraic groups*, Institute for Advanced Study, Princeton, NJ, 1961.

[Wei67] A. Weil, *Basic number theory*, Grundlehren Math. Wiss., vol. 144, Springer-Verlag, New York, 1967.

Index